PUBLICATIONS OF THE ISRAEL ACADEMY
OF SCIENCES AND HUMANITIES

SECTION OF SCIENCES

FAUNA PALAESTINA

ARACHNIDA II: ARANEAE: THOMISIDAE

1. *Tmarus* on stem
(photograph courtesy of P. Amitai)

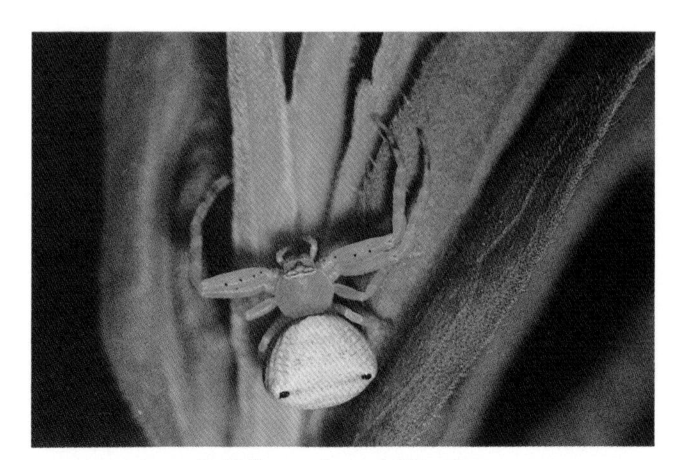

2. Yellow coloured *Thomisus*
(photograph courtesy of N. Ross)

3. *Xysticus* guarding egg sac
(photograph courtesy of A. Shoob)

FAUNA PALAESTINA • ARACHNIDA II

ARANEAE : THOMISIDAE

by

GERSHOM LEVY

Jerusalem 1985

The Israel Academy of Sciences and Humanities

Author's Address:
Department of Zoology
The Hebrew University of Jerusalem
91904 Jerusalem, Israel

ISBN 965-208-058-6

Printed in Israel
at Keterpress, Jerusalem

CONTENTS

PREFACE 1

ACKNOWLEDGEMENTS 2

INTRODUCTION 3

SPIDER MORPHOLOGY 6

CRAB-SPIDERS (THOMISIDAE)
General Biology and Ecology 13

SYSTEMATIC PART
FAMILY THOMISIDAE 17
Key to the Genera of Thomisidae in Israel 17
Genus Monaeses 18
Genus Tmarus 22
Genus Firmicus 31
Genus Thomisus 35
Genus Runcinia 43
Genus Heriaeus 47
Genus Synaema 53
Genus Oxyptila 60
Genus Xysticus 71

APPENDIX
List of Emendations 108

REFERENCES 110

INDEX 111

MAP: GEOGRAPHICAL AREAS IN ISRAEL AND SINAI *end of book*

COLOUR PLATE *opposite title page*

PREFACE

Species of thomisid spiders have been described from Israel and adjacent countries at various times, mostly, however, over a hundred years ago. Some were collected sporadically by travelers, others by a few visiting scientists, the material being dispersed throughout Europe. Implementing a revision thus called for great efforts, extending over a long period of time: first locating the specimens, then enlisting cooperation, and eventually the receipt, examination and return of each specimen.

I examined specimens of all previously known species from our area, presently kept in the following institutions: O. Pickard-Cambridge collection in Oxford University, E. Festa collection (determined by P. Pavesi) in the Zoological Museum of Torino University, V. Kulczyński collection in the Polish Academy, Warsaw, and J. Aharoni's collection (determined by E. Strand) in the Senckenberg Museum, Frankfurt-am-Main. In addition, the collections of the Muséum National d'Histoire Naturelle, Paris (mainly E. Simon's collection) and that of the British Museum (Natural History), London, were studied for relevant material.

All species, except for *Oxyptila clavigera* (O. P.-Cambridge, 1872), previously described from this region were rediscovered in the field. Specimens of these, along with the others dealt with here, are deposited in the collections of The Hebrew University of Jerusalem.

The descriptions given are in abbreviated form, chiefly relating to genital characteristics important for identification. The terminology used is explained in the morphological section and the labelled drawings (Figs. 1–10). The male and female genitalia of each species are depicted, showing the ventral and lateral aspects of the male pedipalp and the external and internal (dorsal) view of the female epigynum.

Measurements are from preserved material and were taken, whenever possible, of specimens from localities throughout the geographical range. The numbers in parentheses which appear after the distribution records refer to areas on the map at the end of the book. Full documentation and more detailed record information are to be found in papers by the author, published by the *Israel Journal of Zoology* (1973, 1975 and 1976).

ACKNOWLEDGEMENTS

For the loan of specimens and types I am deeply indebted to Prof. M. Vachon and Dr M. Hubert, Muséum National d'Histoire Naturelle, Paris; Dr G. H. Sheals, Mr K. H. Hyatt and the late Mr D. J. Clark, British Museum (Natural History), London; Prof. G. C. Varley, Dr M. W. R. de V. Graham and Mr D. M. Ackland, Hope Entomological Collections, University Museum, Oxford; Prof. O. Kraus, formerly of the Senckenbergische Naturforschende Gesellschaft, Frankfurt a.M.; Dr W. Starega, Zoological Institute, Polish Academy of Sciences, Warsaw; Prof. S. Karol, Department of Zoology, Ankara University; Prof. U. Parenti and Dr O. Elter, Museo ed Istituto di Zoologia Sistematica, Torino University.

Sincere thanks for drawing most of the illustrations are due to Miss S. Halbreich and the late Mr G. Tsabar. We are grateful to the Weizmann Science Press of Israel, publishers of the *Israel Journal of Zoology*, for permission to reproduce figures from papers by G. Levy (1973, 22: 107–141; 1975, 24: 155–175 and 1976, 25: 1–37).

This study was supported by the Fauna Palaestina Committee of the Israel Academy of Sciences and Humanities.

INTRODUCTION

About 30,000 species of spiders have been named so far, representing only a part of all living spiders. With the possible exception of the mites, spiders constitute the largest group of arachnids. They are found in all terrestrial environments, occupying virtually every conceivable ecological niche. Many species live on marine and freshwater shores, some below the high-tide line, presumably hiding in air pockets under rocks. Only one spider, though breathing atmospheric air, is entirely aquatic. Their great diversity is reflected in a wide range of life styles, behaviour, and morphological and physiological adaptations. Quantitative examinations of soil and litter fauna reveal surprisingly dense populations of spiders. All spiders are obligate carnivores and are considered among the most important insect predators in nature. The spiders thus have a significant impact on natural ecosystems.

Spiders may be divided into two broad groups based on their hunting strategies: those with generally sedentary habits, which employ some form of silken snare suspended in a more or less permanent station; and spiders that forgo the use of a snare and range over the substratum in search of prey.

Silk is the characteristic product of spiders. Most spiders use silk to build a retreat and to form a sac enclosing their eggs; many also construct a web to snare prey. Spider silk is considered to have the highest tensile strength of any known natural fibre. This tensile strength, combined with the silk's great elasticity, has been exploited remarkably by the spiders. There are about six different types of silk glands. These are considered to produce distinct types of silk. All spider silks are proteins, but the chemical and physical properties of the different silks produced by the same spider vary considerably. The spider can manipulate the various silks, producing the appropriate kind for the function it is to perform.

The web structure varies enormously in design, complexity and function. Except for minor individual modifications, the form of web built by a spider is fixed. The web structure is likely to conform to a certain set of characteristics of a potential prey. Thus appropriate size and mode of movement through space and on the surface make certain prey vulnerable to ensnarement by certain webs. The individual spider has some latitude, within the range of its physiological needs and tolerance, in the selection of a site for its web, a choice which may strongly influence the quantities and forms of prey to be captured.

Unlike web builders, wandering spiders may exercise greater choice of preference with respect to forms of prey. Apparently, searching for prey at random is the general pattern of many hunting spiders which overtake their victims by active pursuit. On sighting a potential prey, they charge forward and seize it with their forelegs and mouthparts. Others creep close to the prey and then pounce on it with lightning speed. Some species live near water and can run across the water surface or descend below it by walking down banks or plant stems in order to prey there on aquatic insect larvae. Ambush is another common hunting strategy of many species. Some spiders station themselves on flower corollas in order to ambush visiting insects seeking pollen or nectar, while others cling motionless to stems of plants and grasses, awaiting the approach of a victim.

Ecologically, spider communities show a certain vertical stratification: the ground-level community and the plant-living community, namely field and canopy layers. Generally, all hunting strategies are encountered in each of the strata; species sometimes move between the layers.

Most spiders possess eight eyes, a pair of principal eyes and three pairs of auxiliary eyes. All eyes are simple ocelli with a single lens. However, some species have good eyesight, being able to perceive prey from a distance of several centimetres. The auxiliary eyes are suited apparently for the detection of movements, and the spider then orients its body so as to centre the principal eyes on the source of movement. Many short-sighted spiders are nocturnal, their prey being detected by direct contact with the leg tips and then rapidly caught.

All known spiders, with the exception of very few small families, are venomous. The venom is of variable utility to different types of spiders in handling their prey. Some rely entirely on its virulence to subdue their prey, while others, including those lacking venom glands, immobilize their prey with thick bands of silk. The venom of many spiders is lethal to insects. Relatively few species of spiders are known to produce venoms which have neurotoxic fractions of sufficient potency to affect vertebrates and man; some venoms are known to produce necrotic lesions.

Spiders are incapable of accepting any solid food and rapidly acting digestive fluids are pumped into the body of the prey. The softened inner contents of the prey are then sucked into the spider gut by powerful pumps. The empty exoskeleton of the prey remains in perfect condition and may be identified with relative ease. Some spiders, which have teeth, crush and chew up the prey, leaving only a compressed ball of unrecognizable exoskeleton fragments.

The sexes are separate in all spiders and, with the exception of very few records of parthenogenesis, all reproduce sexually. Usually males are smaller (by a factor of up to ten) and have longer legs than the females. The small males go through fewer moults than do the females and mature much earlier. Sexual differentiation only becomes apparent in the later stages of development. The tips of the males' pedipalps become swollen and after the final moult contain complex copulatory organs. Before searching for a female, the male deposits a drop of sperm on a special web, then dips a palp tip into it and fills the palpal organ with semen (sperm induction). The search of the male for a mate may be associated with a sex pheromone emitted by the female. Courtship by the adult male varies greatly in the different families; it may involve long and complex displays. Subsequent to mating, a male is only infrequently treated as prey by the female; males usually die very soon thereafter.

Spiders lay eggs in batches usually wrapped in a silken sac, often made of layers of various types of silk; some bind the eggs together with a scant network of simple silken strands. The number of eggs per sac varies from a few in small spider species to several hundred in the large spiders. The egg sac may be suspended in the web, carried in the mouthparts, or attached posteriorly to the mothers' spinnerets. Many species guard their eggs until hatching; some also guard the young until they disperse.

Spiders disperse by methods common for land arthropods, also making extensive use of aerial bridges of silk. In certain families the spiderlings are known to disperse by a process called ballooning. The young climb to the tip of some object, eject silk which is carried aloft by the wind, and then release their grip and become airborne.

Spiderlings usually moult at least once before leaving the egg sac. The number of moults and the rate of growth may be fairly constant in some species or vary considerably in others, even in offspring of the same batch. Female spiders of a few families considered to be primitive moult after maturity.

Spiders are able to curtail their growth and rate of development radically during extended periods of prey scarcity and have the complementary ability to take advantage of even brief periods of high prey densities in order to accelerate development and increase growth. The spiders' functional response to changes in prey density may thus have profound effects. The potential of spiders as agents of biological control should, therefore, be considered of significant economic importance.

Spiders are collected by turning stones, sweeping shrubs and herbs, looking under tree bark and watching for holes in the ground. Some species are nocturnal and are observed best at night with the aid of a lamp.

Spiders can be chased into a glass vial or lifted with forceps, but, being soft-bodied, they should be handled with care. In rearing, each spider must be kept in a separate container to prevent cannibalism. Although many spiders do not need food for days, a few drops of water should, however, be provided regularly. Flies, mealworms or young grasshoppers are generally used for feeding.

Spiders are preserved in 70% alcohol and are examined while completely submerged, never dry. The male palp, usually the left one, is studied from the various angles useful for species recognition. The genital aperture of the female, after being examined from the outside, has to be dissected carefully for the study of the internal structures; if possible, the dissected part should not be removed from the body of the specimen.

SPIDER MORPHOLOGY

The spider body consists of two parts, the anterior *prosoma* (= cephalothorax) and the posterior *opisthosoma* (Figs. 1, 2). The two portions are connected by a narrow stalk, the pedicel, enabling free movement of each portion in all directions. The prosoma combines the head and thorax functions; it contains the highly condensed nervous system and bears the eyes, mouthparts, pedipalps and walking legs. The opisthosoma bears the openings of the respiratory, reproductive and digestive systems, and also the spinnerets.

Prosoma: It is covered above by an unsegmented *carapace* (Fig. 1) and below by a large plate, the *sternum* (Fig. 2). Attached to the anterior edge of the sternum, usually delimited by a groove, is a median platelet serving as a *labium* (Figs. 2, 4). The sternum is often indented opposite the coxa of each leg. The *eyes* are variously distributed over the carapace (Figs. 1, 3). Their disposition is characteristic in different families and is used extensively in taxonomy. They are often arranged in two

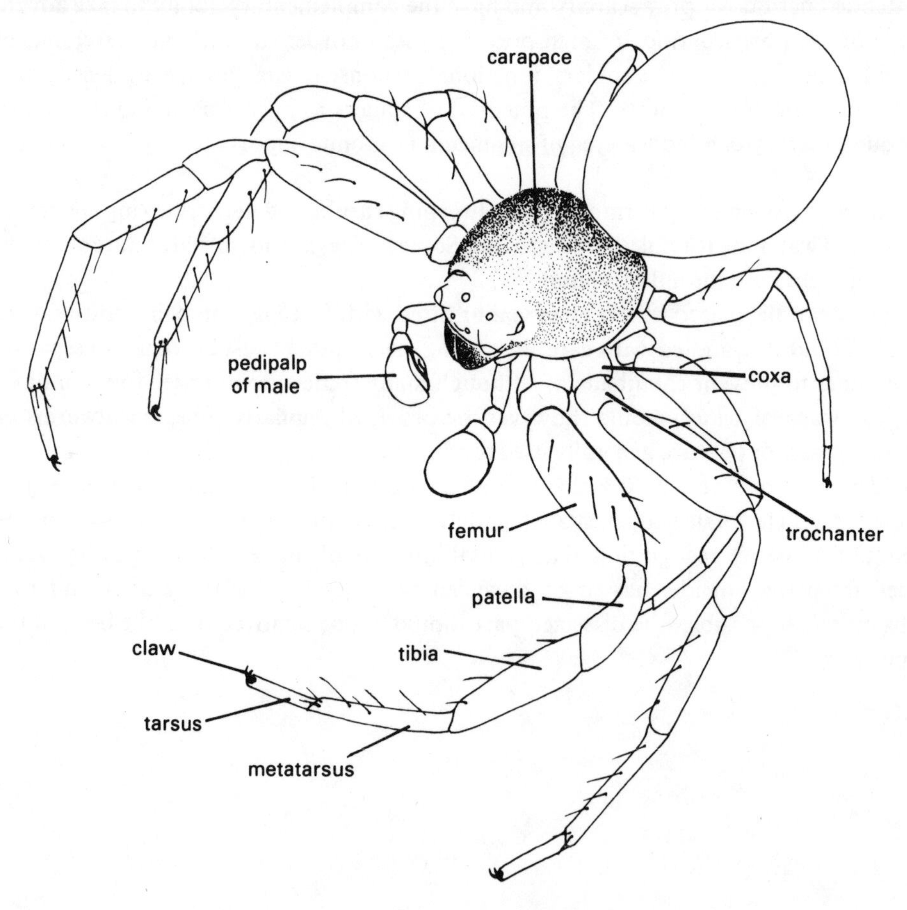

Fig. 1: Male spider, dorsal view

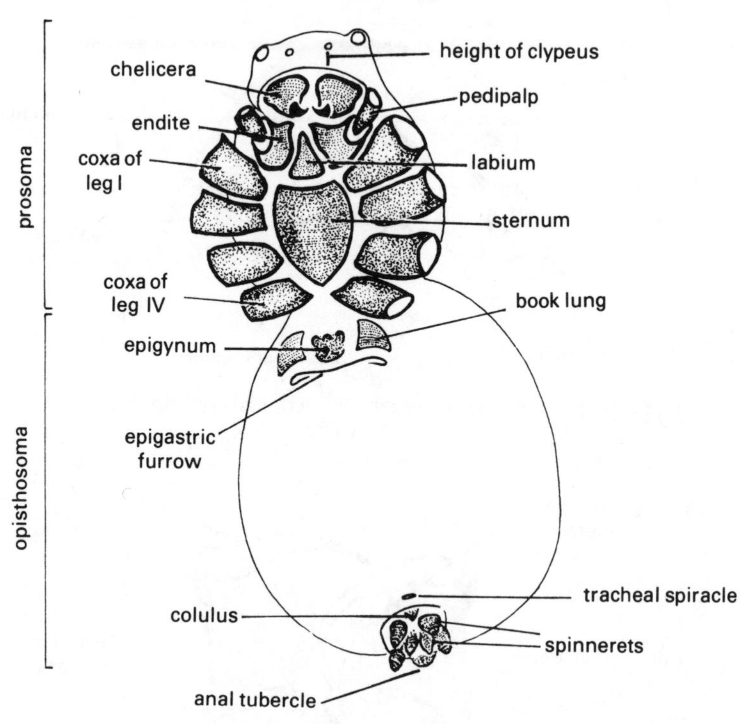

Fig. 2: Female spider, ventral surface
illustrating main external features (legs omitted)

rows, the *anterior row* and the *posterior row*, the inner two eyes of each row being the *medians*, the outer eyes — the *laterals* (Fig. 3). This terminology is retained throughout even though in some spiders the specific term does not actually describe the location of a particular pair of eyes. The rows of eyes are frequently curved. When the lateral eyes of a row are nearer to the front end of the carapace than are the median eyes, the row is termed *procurved*, and the opposite situation is termed *recurved* (Fig. 3). In specifying the curvature it is assumed that the eyes are viewed from vertically above; the actual curve of the anterior row is checked from directly in front. The area enclosed by the anterior-median and posterior-median eyes is termed the *median ocular quadrangle* and is useful taxonomically (Fig. 3). When both dark and light-coloured eyes are present in the same spider they are referred to as *heterogeneous*, whereas if the eyes are all alike they are called *homogeneous.* The region between the anterior eyes and the front margin of the carapace is termed the *clypeus.* The *height* of the clypeus refers to the distance from the front edge of the carapace to the eyes nearest that edge (Fig. 2).

Mouthparts: A pair of *chelicerae* are positioned above the mouth. Each chelicera consists of a stout basal segment and a distal fang (Fig. 4). The fang folds onto the basal segment along a groove whose edges are frequently provided with rows of small teeth. The teeth are often of importance in identification. Those on the outer margin

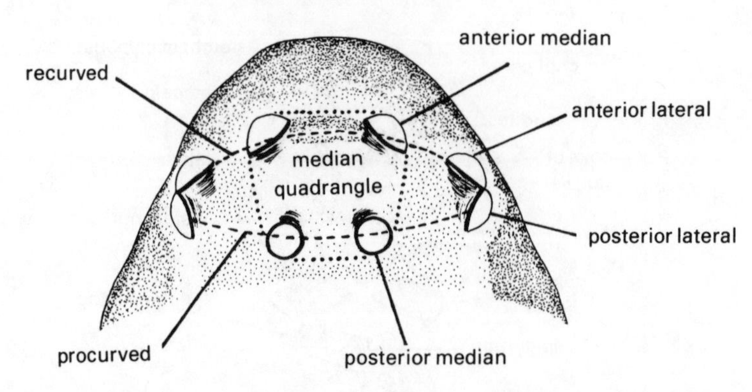

Fig. 3: Diagram of eyes' arrangement on dorsal, anterior part of carapace

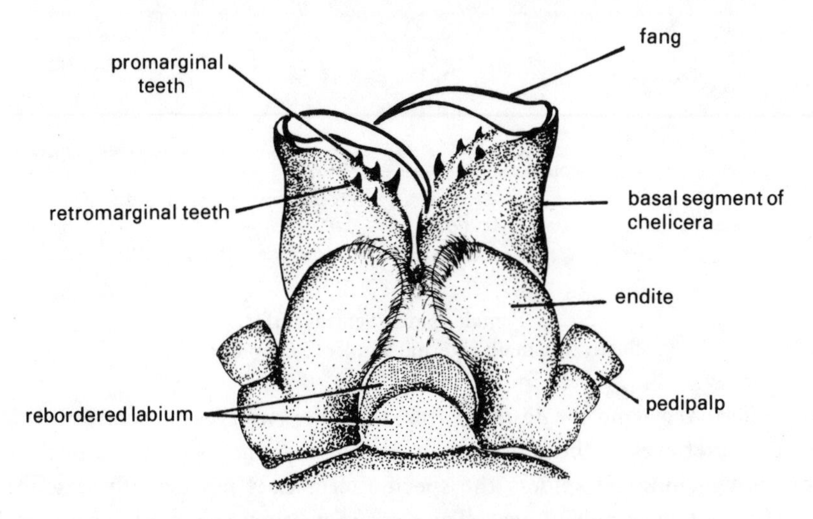

Fig. 4: Mouthparts of a spider, ventral view

(= anterior or upper row) are termed *promarginal* and those on the inner margin — *retromarginal* (Fig. 4); the latter may be visible from below. Close to the tip of the fang is the opening of the poison gland. This gland either lies entirely in the chelicerae or, as in most spiders, extends into the prosoma.

Pedipalpi: Situated behind the chelicerae, the pedipalps, usually referred to as *palps*, are leg-like in females, or modified in mature males to contain the copulatory organs (Fig. 1). In most spiders their coxae are greatly enlarged on the inner side to form flattened *endites* enclosing the preoral cavity (Figs. 2, 4). In the male palp, often the tibia, sometimes also the patella and even the femur, may be enlarged and bear *apophyses* and special spines which are of great taxonomic importance (Figs 5, 6). In the majority of spiders the distal part of the male palp, the *cymbium*, is hollowed out to accommodate the functional components of the copulatory organ. The palpal organ of the male palp stores the sperm derived from the genital orifice situated below the anterior part of the opisthosoma, and is used to transfer the sperm into the

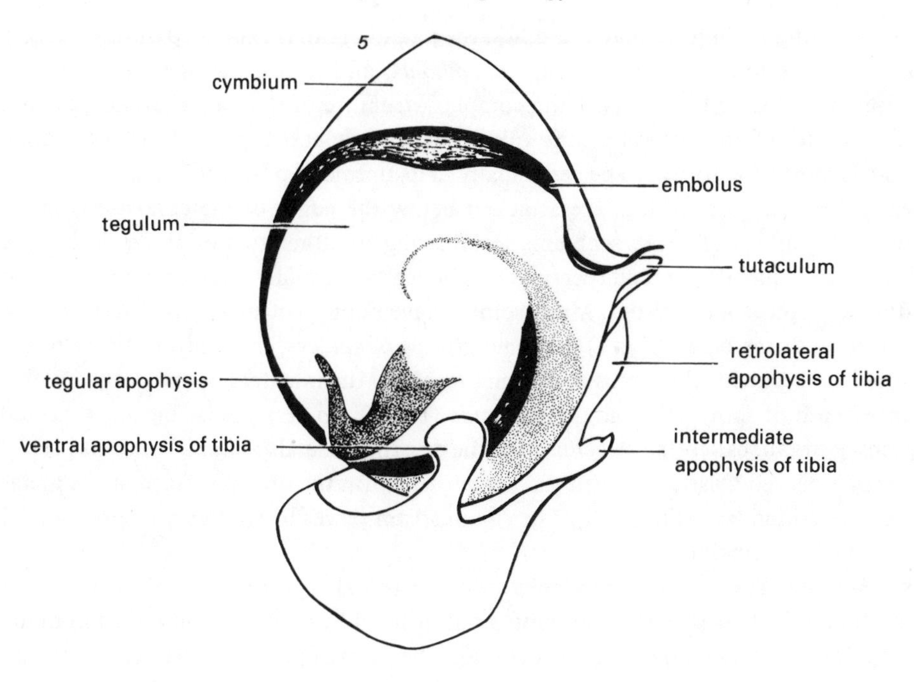

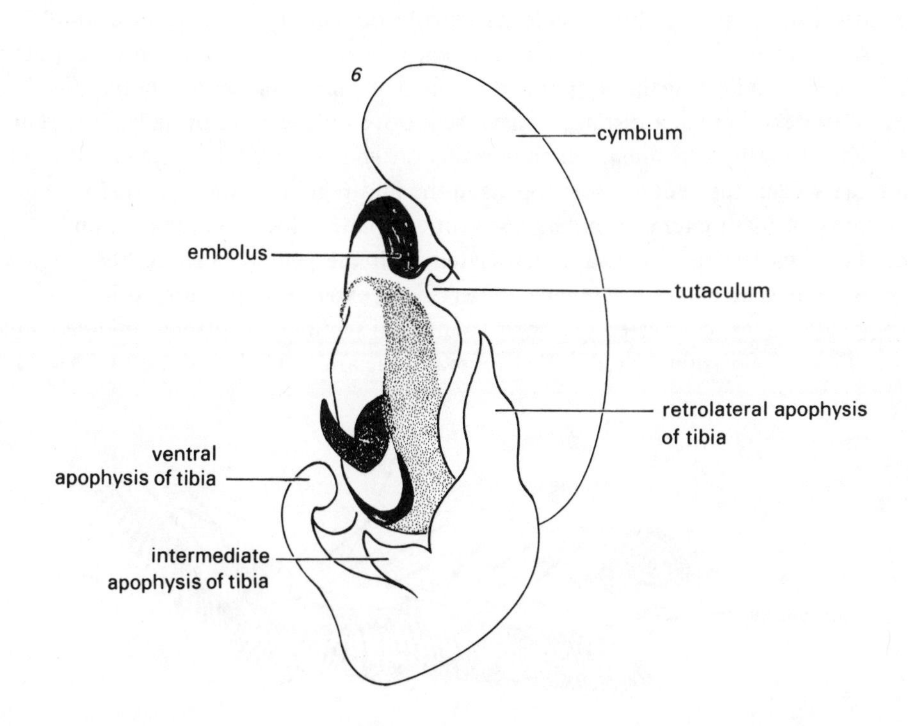

Figs. 5–6: Diagram of male palpal organ; left palpus
5. ventral view; 6. lateral view

female during mating. In many families, the palpal organ is expanded during mating, unfolding the intromittent portion, the *embolus*, and its various sclerites and apophyses (Figs. 5, 6). In the male, the complex structure of the palpal organ provides the character of greatest systematic value, particularly at the species level. The palpal organ is illustrated for each species, usually in its unfolded position.

Legs: The four pairs of legs are attached below the edges of the carapace (Fig. 1). Each leg consists of seven segments terminating in either two or three claws. The claws are frequently toothed (pectinate). The legs are usually covered with hairs and often bear spines and bristles. Many spiders have bundles of strong hairs termed *claw tufts* beneath the claws (Fig. 7). In some groups of spiders, the ventral surface of the leg's distal segment, the tarsus and sometimes also the metatarsus, are clothed with a dense brush of short, stiff hairs, the *scopula* (Fig. 7). Spiders possessing a specialized spinning organ, called the *cribellum*, on the opisthosoma also have a comb-like series of small, curved bristles, termed the *calamistrum*, on the dorsal surface of the metatarsi of the hind legs (Figs. 8, 9). The calamistrum serves to card very fine threads of silk out of the cribellum.

Opisthosoma: The opisthosoma lacks visible external segmentation in all spiders of the Middle East; it is most commonly oval in form but it may be modified in many ways. Some of these modifications may take the form of protuberances or a long tail. Many spiders have a coloured pattern on their back and venter, but it may vary greatly within species; bright colours usually do not persist long in alcohol. The dorsal surface of the opisthosoma frequently bears a series of depressed spots of varying size, indicating the points of attachment of internal muscles to the body wall.

In all spiders there is a distinct transverse groove on the anterior half of the ventral surface of the opisthosoma, which is called the *epigastric furrow* (Figs. 2, 10). In the mature spider, the genital orifice opens in the mid-region of this furrow. In the great majority of mature female spiders a sclerotized plate is located in the region in front of the epigastric furrow and is associated with the genital opening, the *epigynum* (Figs. 2, 10). Most mature female spiders have external copulatory orifices on the epigynal surface, connected via ducts to internal seminal receptacles, the *spermathecae* (Fig. 10). Sperm is stored in the spermathecae and passed via ducts to the oviduct

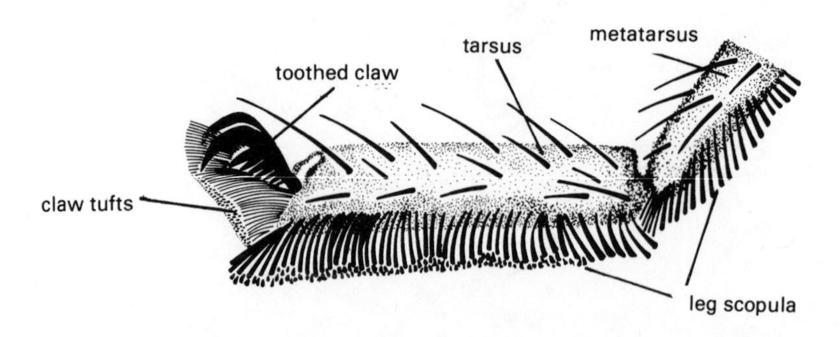

Fig. 7: Terminal joints of walking leg with scopula and claw tufts

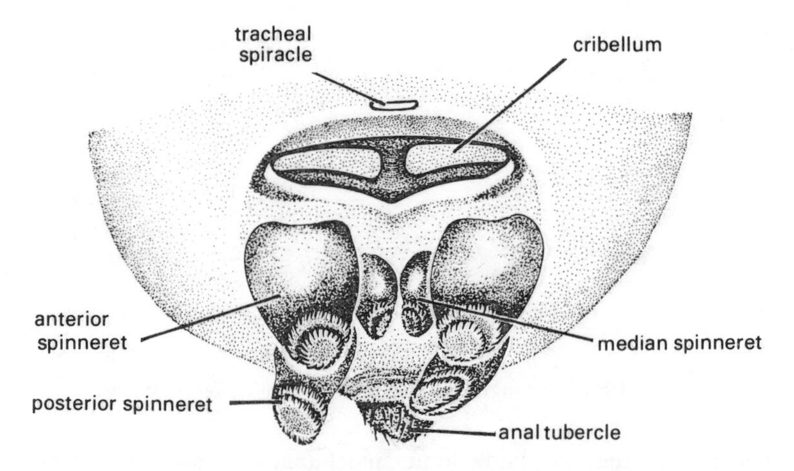

Fig. 8: Anal part of spider showing cribellum and spinnerets

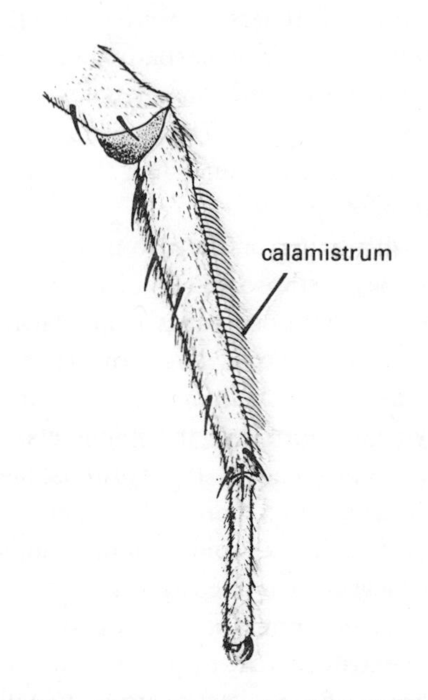

Fig. 9: Fourth leg, portion showing bristles on metatarsus forming the calamistrum

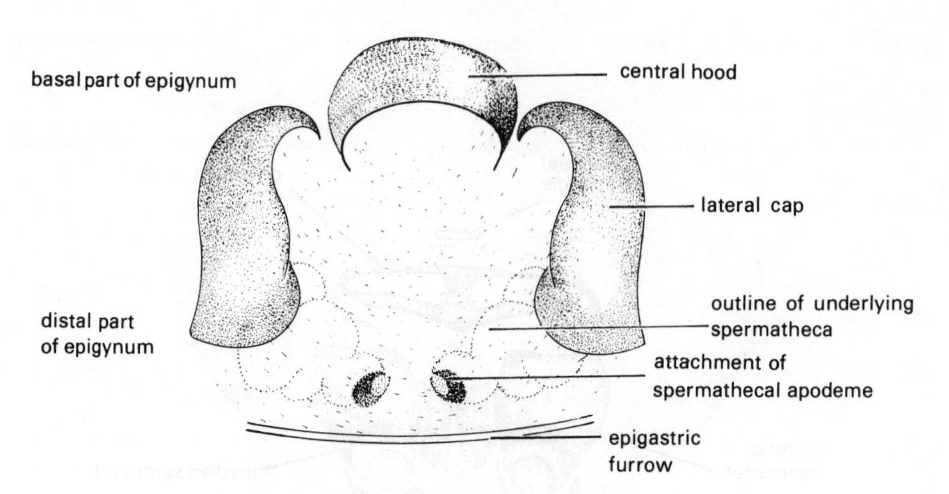

Fig. 10: Diagram of epigynal plate of female, ventral view

to fertilize eggs as they are laid. Some spider families have no separate, external copulatory pores and the spermathecae open only into the oviduct; the intromittent organ of the male palpus in these spiders is inserted directly into the genital orifice. The internal spermathecal organs may be partially discernible through the epigynal integument; dark circles, indicating attachment points of internal apodemes, are sometimes visible on the outside (Fig. 10). The epigynum, spermathecae and also the ducts show an infinite variety of form and, like the male palpal organ, are very important in the identification of species.

Spiders may be equipped with either one or two pairs of respiratory organs which open on the ventral side of the opisthosoma (Fig. 2). In the different families, several combinations of lamellate organs ('book lungs') and tracheae of various types and origins are to be found. A pair of book lungs covered by light-coloured plates is situated in front of the epigastric furrow in most spiders, and a single slit of the fused tracheae opens posteriorly near the base of the spinnerets.

At the morphologically posterior end of the opisthosoma is the opening of the digestive system, which is situated on a more or less distinct *anal tubercle* (Figs. 2, 8). Below this are the *spinnerets* which in all spiders of the Middle East may number one, two or three pairs. The spinnerets are jointed appendages — some reduced to a single joint — with a terminal, membranous portion studded with many minute tubes through which liquid silk is extruded. The disposition, relative length and number of joints of the spinnerets are useful in classification. Usually, the medians are the smallest and are often hidden by the others. In a number of families there is a sieve-like plate, the *cribellum*, in front of the spinnerets, from which a special type of silk is carded (Fig. 8). Many of the spiders lacking a cribellum have a vestigial, conical lobe, termed the *colulus*, between the bases of the anterior spinnerets (Fig. 2). Silk glands with external, isolated pores are also found anteriorly on the opisthosoma in many male spiders.

CRAB-SPIDERS (THOMISIDAE)

General Biology and Ecology

Typical thomisid spiders are commonly called crab-spiders as they extend their legs in a crab-like fashion and are able to move forwards, backwards or sideways. For the most part, they have short, wide, considerably flattened bodies. They are all hunting spiders which do not construct a snare but depend mainly on ambush to catch their prey. They wander about freely on the ground and on plants or lurk in crevices, beneath stones and under the bark of trees; some lie appressed to the surface of plants in the open.

Species of *Xysticus* and *Oxyptila* are preeminently ground spiders; their coloration — dull grey, brown, and black — melds with that of the leaves and soil. The ambushing spiders that live on vegetation and in flowers, such as species of *Synaema* or *Thomisus*, are much more brightly coloured than the ground forms. The best-known flower spider of our region is the white, yellow or pink-coloured *Thomisus onustus.* It has the ability to change colour from white to yellow and vice versa, thus matching to some extent the colour of the substratum as seen by the ultraviolet-insensitive eyes of men and predacious birds. Flowers, however, may possess ultraviolet reflection patterns and the spider is potentially detectable by approaching insects that are sensitive to ultraviolet light. While specimens of *Thomisus* may be found on a variety of coloured flowers, a very high percentage occur on white or yellow ones. Generally, in *T. onustus* the change of colour from white to yellow takes about ten days or more, but the reverse process usually requires only three or four days. The rate of change, in whatever direction, is subject to great individual variation. The stimulus for the colour change is conveyed through the eyes; covering of the eyes eliminates any response. Both immature spiderlings and adults of *T. onustus* are capable of changing their colour, unlike *Misumena vatia* which is absent from our region. Local species of *Runcinia* and *Heriaeus* also have the ability to change colour from yellow to green or from green to brown, but only following a moult, not between moults.

Thomisids hiding in flower corollas seem largely to capture day-flying insects, though their reliance on touch rather than sight would appear to make them equally capable of hunting by day and by night. The approach of an insect to the flower on which the spider is poised will trigger adjustments in the spider's position, which are carried out so smoothly that they are barely noticeable. With their robust forelegs and apparently very potent venom, they quickly subdue insects much larger than themselves, including bumblebees, wasps and predacious flies. The thomisids do not crunch their food and consequently the spider retains a certain protective advantage while it sucks the undamaged insect which continues to rest in a natural position on the flower. When a species of *Thomisus* or, especially, *Synaema* resting on a flower is disturbed, it will run beneath a leaf or drop to the ground.

The principal season for mating and laying eggs is spring and early summer. The young females which hatch from these eggs are the mothers of broods in the same season of the following year. The males are distinctly smaller than the females, their sight is very limited and they resort to non-visual courtship. The copulatory postures assumed by the crab-spiders are essentially alike. The male invariably tries to climb upon the female's back, then around to her venter, the resultant position being such that his ventral surface is appressed to the posterior part of the female, heads pointing in the same direction. This position varies slightly according to the particular shape of the female's opisthosoma. In view of the considerable disparity in size between the sexes, the female may sometimes walk around for quite a while with the male clinging to her broad opisthosoma until she finally becomes quiescent. Mating of the typical crab-spiders is a lengthy process which may last more than an hour, during which time the male's palps are in alternate use many times.

The life-span of the sexually mature male is relatively short. During this brief period he may be accepted by one or several females. In captivity males copulate with the same female many times.

Crab-spiders are able to spin at all ages but they make little use of this ability: they produce draglines by means of which they drop to the ground and in order to construct egg sacs. Their egg sacs are much alike in form and texture. They are usually constructed at night, starting with a few silken strands suspended around the crouching female. A discal silk substratum is laid down, the eggs are deposited upon it, and then the roofing is spun over the mass. In *Xysticus, Oxyptila* and *Synaema* the sac, which is white, is usually lenticular in shape, made up of two discs firmly fastened together at the margins. In *Thomisus* more strands are added above the egg sac which is firmly attached to the substratum; the mother guards the eggs. Partially enclosed in a bent leaf or hidden under a stone, she readily takes food while mounting her guard.

In many species of *Xysticus*, the female spider clutches the egg sac with her forelegs and if she is moved, the sac goes with her. Species of *Synaema* build a retreat on twigs of trees; such nests are much more firmly constructed. In the case of crab-spiders the guarding of the egg sac is for the most part a passive role, since the female simply rests on the egg sac or grasps it securely with her legs; she will resist any efforts to separate her from it, but will not attack an interloper.

Thirty to 50 eggs are perhaps the most common number to be laid in a batch by such genera as *Runcinia, Synaema* or *Oxyptila*; but the larger species of *Thomisus* or some *Xysticus* may lay up to 400. Female spiders are known to build a number of egg sacs in succession. In *Thomisus* two to three sacs may be spun at intervals of several weeks, the eggs in the later sacs usually being fewer in number than those in the first brood. Other thomisids, like *Xysticus* or *Oxyptila*, may spin only a single sac. When multiple egg sacs are spun by a single female, all the eggs may be fertile or some of the later ones may be infertile. Fertility is often due to a single insemination; this has been verified in breeding experiments in the laboratory. The semen remains viable for long periods in the female reproductive system. A female will lay eggs and enclose them in an egg sac irrespective of whether fertilization has occurred or not, and will guard them.

The young of *T. onustus* emerge from the egg sac about a month after oviposition. In species of *Xysticus, Synaema* and *Oxyptila*, this period is possibly somewhat shorter.

The rate of growth and the duration of each of the following stadia vary greatly. Males of *T. onustus* reach maturity after three to five moults, while females usually undergo six to eight moults. The sexes also differ distinctly in the amount of food consumed and the dimensions of captured prey; adult males hardly feed at all or, at most, prey on larvae or very small insects, their diet thus equalling that of immature females of the second or third instar. Most of the local crab-spiders mature in the spring, and all maintain an annual cycle, producing one generation a year. Adult males are very short-lived (usually less than a few weeks), while the life-span of adult females is several months.

Distribution: Nearly 40 species of Thomisidae are recognized at present in Israel and Sinai. Some are known from only one of the sexes. Generally, spiders are found singly and when the two sexes are of different appearance, their matching is often problematic. In *Oxyptila, Xysticus* and some other thomisid genera with several species inhabiting the same faunal area, the two sexes may not be matched with certainty unless substantial evidence has been provided.

Unfortunately, little is known about the spider fauna of adjacent areas, and there is an almost complete lack of recent information. The fauna of the Thomisidae in Israel consists mainly of Palaearctic (Mediterranean) species. Some are eremic (Palaeo-eremic) and one or two are probably Ethiopian. Several species are found only along the Jordan Rift. Owing to the sparse information available, it is not yet possible to ascertain whether these represent relicts of an ancient tropical fauna, an influx of Ethiopian elements, or are Palaeo-eremic species which have adapted themselves to the special environmental conditions of oases in the Rift Valley.

SYSTEMATIC PART

Family THOMISIDAE Sundevall, 1833

Conspectus Arachnidum, Londini Gothorum, p. 27

Diagnosis: Thomisids are small to medium-sized spiders, usually not exceeding 10–12 mm in body length; the largest are some tropical species, attaining a length of 20 mm. Legs extending sideways in laterigrade position. First two pairs of legs stout and much longer than posterior pairs; legs without scopulae; tarsi bear two toothed claws; claw tufts absent or composed of simple hairs.

Eight small eyes, homogeneous, dark, and arranged in two usually recurved rows; lateral eyes usually raised on tubercles. Respiratory system consists of one pair of book lungs and a single medial tracheal opening situated near the spinnerets. Three pairs of spinnerets, the anterior pair close together; colulus in form of a group of short hairs; cribellum and calamistrum absent.

Only one subfamily, the Misumeninae, with diagnosis as above, occurs in our region; other subfamilies are known mostly from the tropics. The Misumeninae include nearly 80 genera with more than 1200 species widely distributed in all the faunal regions of the world (Roewer, 1954). Nine genera have been recorded in Israel, about the same number as in the whole of the United States north of Mexico or in Central Europe; France, with an influx of African species, has about a dozen. The genera *Xysticus, Oxyptila, Synaema* and *Tmarus* are distributed in the New and the Old World; however, unlike certain other spider families, none of the thomisid species of Israel is also found in the Americas.

Key to the Genera of Thomisidae in Israel

1. Carapace very flat and thin; lateral eyes borne on top of low tubercles (Figs. 33–35). — **Firmicus** Simon
– Carapace rather high or, when partly flattened dorsally, lateral eyes borne on sides of tubercles, not on top — 2
2. Carapace distinctly longer than wide (Fig. 11); opisthosoma slender with posterior part very elongated, extending far beyond spinnerets (Fig. 14). — **Monaeses** Thorell
– Carapace about as long as wide; opisthosoma shaped otherwise — 3
3. Clypeus markedly sloping (Fig. 20); opisthosoma usually with a dorsocaudal tubercle (Figs. 22, 23). — **Tmarus** Simon
– Clypeus vertical; opisthosoma without a caudal tubercle — 4
4. Lateral eyes borne on a common tubercle — 5
– Lateral eyes borne on separate tubercles — 6
5. Tubercles of lateral eyes large, connate and markedly projecting (Fig. 41); opisthosoma triangular, almost truncated and very broad posteriorly (Fig. 42). — **Thomisus** Walckenaer
– Tubercles of lateral eyes low, protruding sideways (Fig. 56); opisthosoma oval, slightly rounded posteriorly (Fig. 55). — **Runcinia** Simon

6. Very hairy spiders, body covered with long, thick bristles sometimes borne on small tubercles (Figs. 61, 62). **Heriaeus** Simon
– Surface of body smooth or sparsely covered with hairs and bristles 7
7. Median ocular quadrangle wider than long, narrow anteriorly and broader posteriorly (Figs. 72, 73). Spiders with brilliant colours (Fig. 71). **Synaema** Simon
– Median quadrangle usually longer than wide or nearly square. Spiders usually with dull yellowish-brown or dark coloration (Figs. 86, 106). 8
8. Anterior-median eyes usually closer to each other than to posterior-lateral eyes; median ocular quadrangle usually longer than wide (Figs. 87, 88). Body covered by bristles, partly clavate or spatulate. **Oxyptila** Simon
– Anterior-median eyes usually farther from each other than from posterior-lateral eyes; median quadrangle nearly square (Figs. 107, 108). Body covered with thin or thick, pointed bristles; clavate bristles absent. **Xysticus** C. L. Koch

Genus MONAESES Thorell, 1869

Nova Acta R. Soc. Scient. upsal., (3) 7 : 37

Figs. 11–14

Monastes Lucas, 1846, Histoire naturelle des Animaux articulés, in: *Exploration scientifique de l'Algérie, Zool.*, 1: 192 (preoccupied).

Type Species: *Monastes paradoxus* Lucas, 1846.

Elongated spiders. Carapace rather high and distinctly longer than wide (Figs. 11, 12). Ocular area elevated and eyes borne on high tubercles clearly separated from each other; lateral eyes distinctly larger than median eyes; tubercles of posterior-lateral eyes much larger than anterior-lateral ocular tubercles; anterior eyes more or less equidistant; posterior-median eyes distinctly farther from each other than from posterior-lateral eyes (Fig. 13). Chelicerae relatively large, directed downwards and obliquely forwards (Fig. 12). Legs thick and very long; two anterior pairs distinctly longer than posterior pairs; claws with large, distinct denticles. Opisthosoma long and slender, dorsally flat, with sides running parallel along median part; posterior part cone-shaped, very elongated, extending far beyond spinnerets, surrounded by conspicuous folds with bristles; posterior part of opisthosoma usually inclined slightly upwards, but occasionally also inclined slightly downwards (Fig. 14).
Monaeses species inhabit dense vegetation, usually near water. They move very slowly and catch their prey by ambush. When not moving, they cling to stems of plants, with their body and legs stretched along the stem's axis, and are thus inconspicuous. Usually these spiders are captured with sweep nets.
About two dozen species, mainly in Africa. One species in Europe and a few in East and South Asia, Australia and Guiana. One species in Israel.

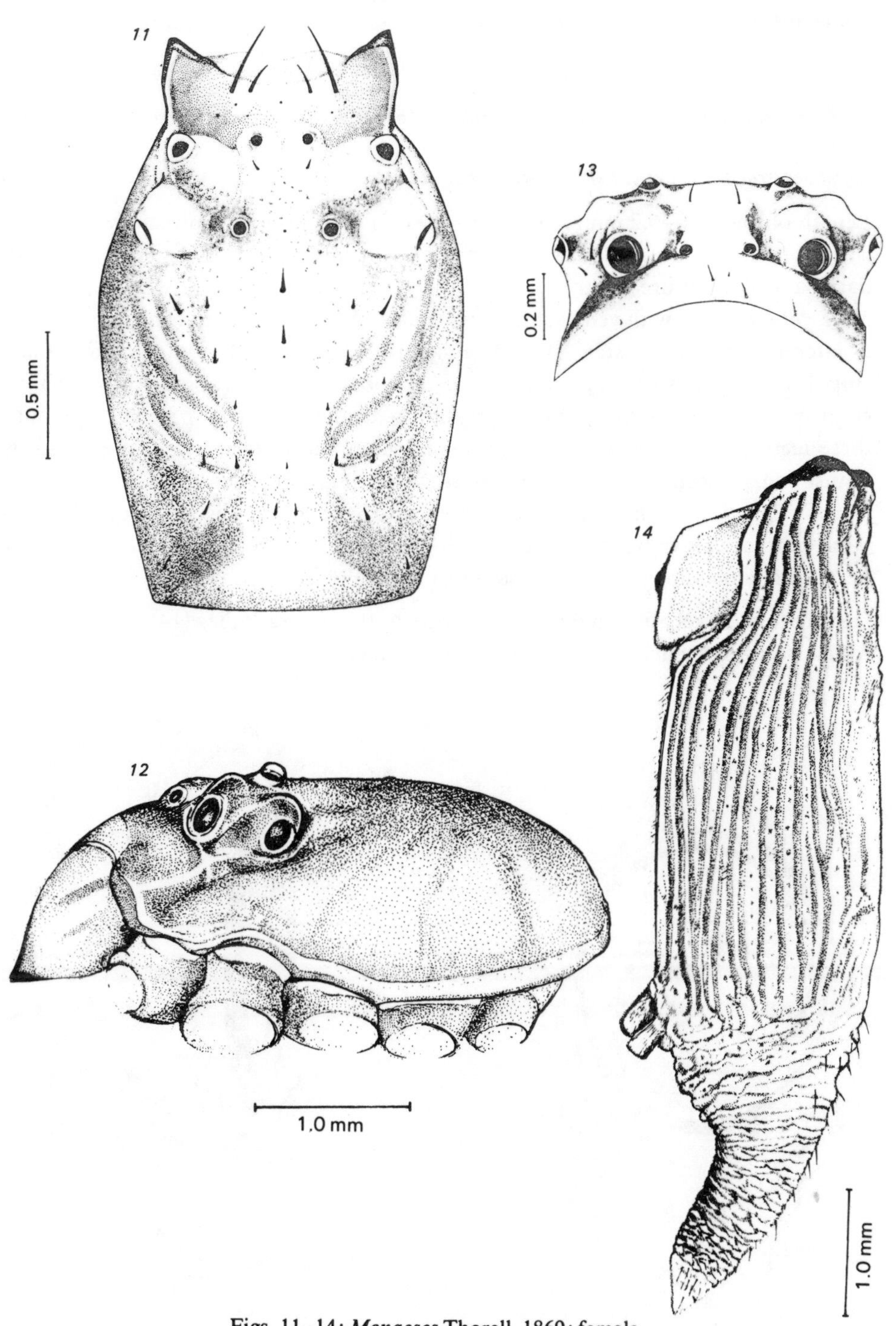

Figs. 11–14: *Monaeses* Thorell, 1869; female
11. carapace, dorsal view; 12. prosoma, lateral view;
13. carapace, frontal view; 14. opisthosoma, lateral view

Monaeses israeliensis Levy, 1973
Figs. 15–18

Monaeses paradoxus (Lucas, 1846). Strand, 1915, *Arch. Naturgesch.*, 81: 147 (misdetermination).
Monaeses israeliensis Levy, 1973, *Israel J. Zool.*, 22: 111.

Length of male 4.0–6.2 mm, female 8.0–9.0 mm. Coloration yellowish-grey or whitish-grey. Opisthosoma slightly darker than prosoma; venter of opisthosoma with dark band along entire length.
Male Palpus: Tibia with ventral and retrolateral apophyses. Ventral apophysis slightly laterally positioned, stem at middle slightly bent; central part of stem swollen (Fig. 16); apophysis terminating in rounded knob. Retrolateral apophysis with short, stout stem and with a groove-like depression ventrally (Fig. 15); tip stellate, split into three teeth (Figs. 15, 16). Tutaculum of palpus broad at base, narrowing towards obtuse end, projecting close to end of retrolateral tibial apophysis (Figs. 15, 16).
Female Epigynum: Dark, sclerotized, crescent-shaped area slightly protruding on basal part of epigynum, forming a hood-like protuberance overhanging a shallow hexagonal depression (Fig. 17; dorsal-inner view, Fig. 18).

Distribution, Israel : Throughout the country to the 'Arava Valley (14).
Adults occur in the field mainly from April to June.

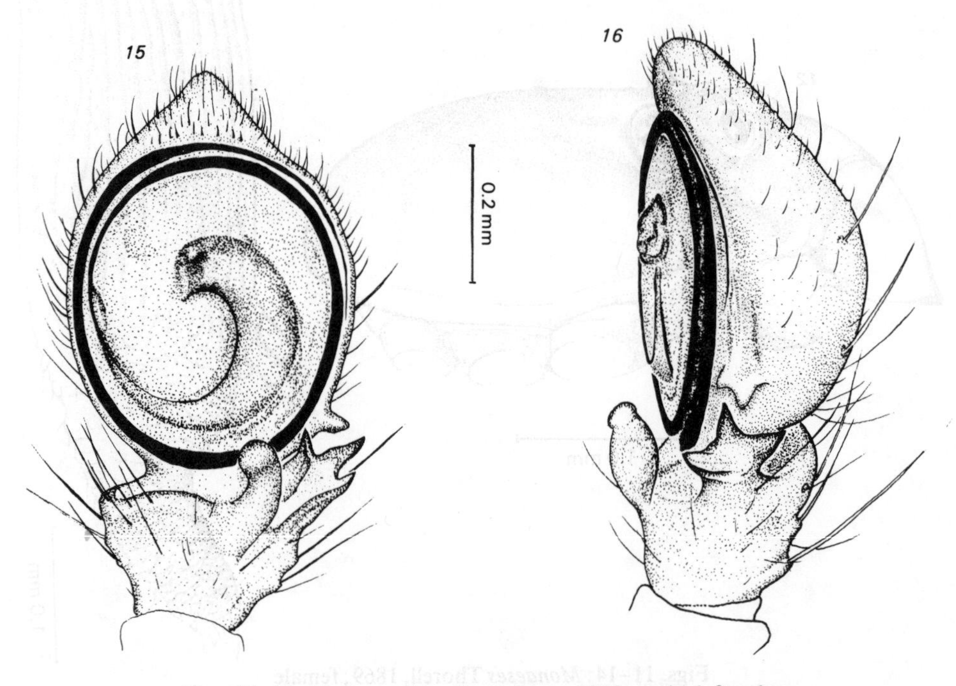

Figs. 15–16: *Monaeses israeliensis* Levy, 1973; male, left palpus
15. ventral view; 16. lateral view

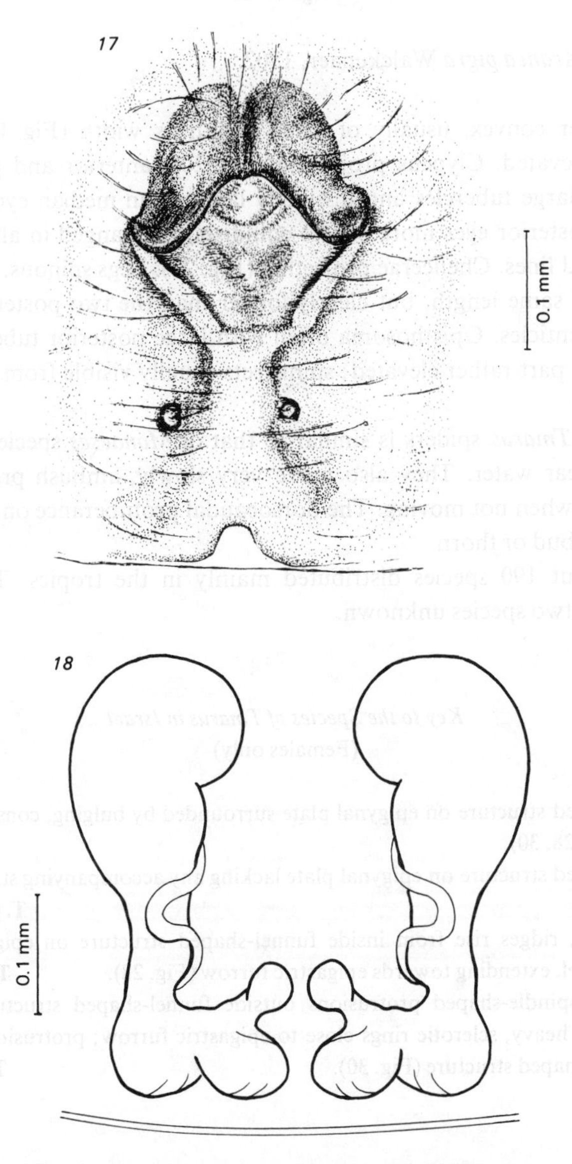

Figs. 17–18: *Monaeses israeliensis* Levy, 1973; female
17. epigynum; 18. spermathecae, inner view

Genus TMARUS Simon, 1875

Les Arachnides de France, 2: 259

Figs. 19–23

Type Species : *Aranea pigra* Walckenaer, 1802.

Carapace rather convex, usually of same length as width (Fig. 19). Ocular area considerably elevated. Clypeus projecting (Fig. 20); anterior and posterior lateral eyes borne on large tubercles and distinctly larger than median eyes (Figs. 20, 21); anterior and posterior eyes more or less equidistant, arranged in almost straight or slightly recurved lines. Chelicerae projecting (Fig. 20). Legs spinous, the two anterior pairs almost of same length, but clearly longer than the two posterior pairs; claws with distinct denticles. Opisthosoma often with large posterior tubercle dorsally or entire posterior part rather elevated; spinnerets usually visible from above (Figs. 22, 23).

The habitat of *Tmarus* spiders is similar to that of *Monaeses* species: dense vegetation, usually near water. They also move very slowly, ambush prey, and cling to stems of plants when not moving. The dorsocaudal protuberance on their opisthosoma resembles a bud or thorn.

Altogether about 190 species distributed mainly in the tropics. Three species in Israel; males of two species unknown.

Key to the Species of Tmarus in Israel

(Females only)

1. Funnel-shaped structure on epigynal plate surrounded by bulging, conspicuous bands or ridges (Figs. 28, 30) — 2
   – Funnel-shaped structure on epigynal plate lacking any accompanying structures (Fig. 26). — **T. piochardi** (Simon)
2. Two bulging ridges rise from inside funnel-shaped structure on epigynum, recurving outside funnel, extending towards epigastric furrow (Fig. 28). — **T. yerohamus** Levy
   – Two dark, spindle-shaped protrusions outside funnel-shaped structure on epigynum project from heavy, sclerotic rings close to epigastric furrow; protrusions do not extend into funnel-shaped structure (Fig. 30). — **T. hazevensis** Levy

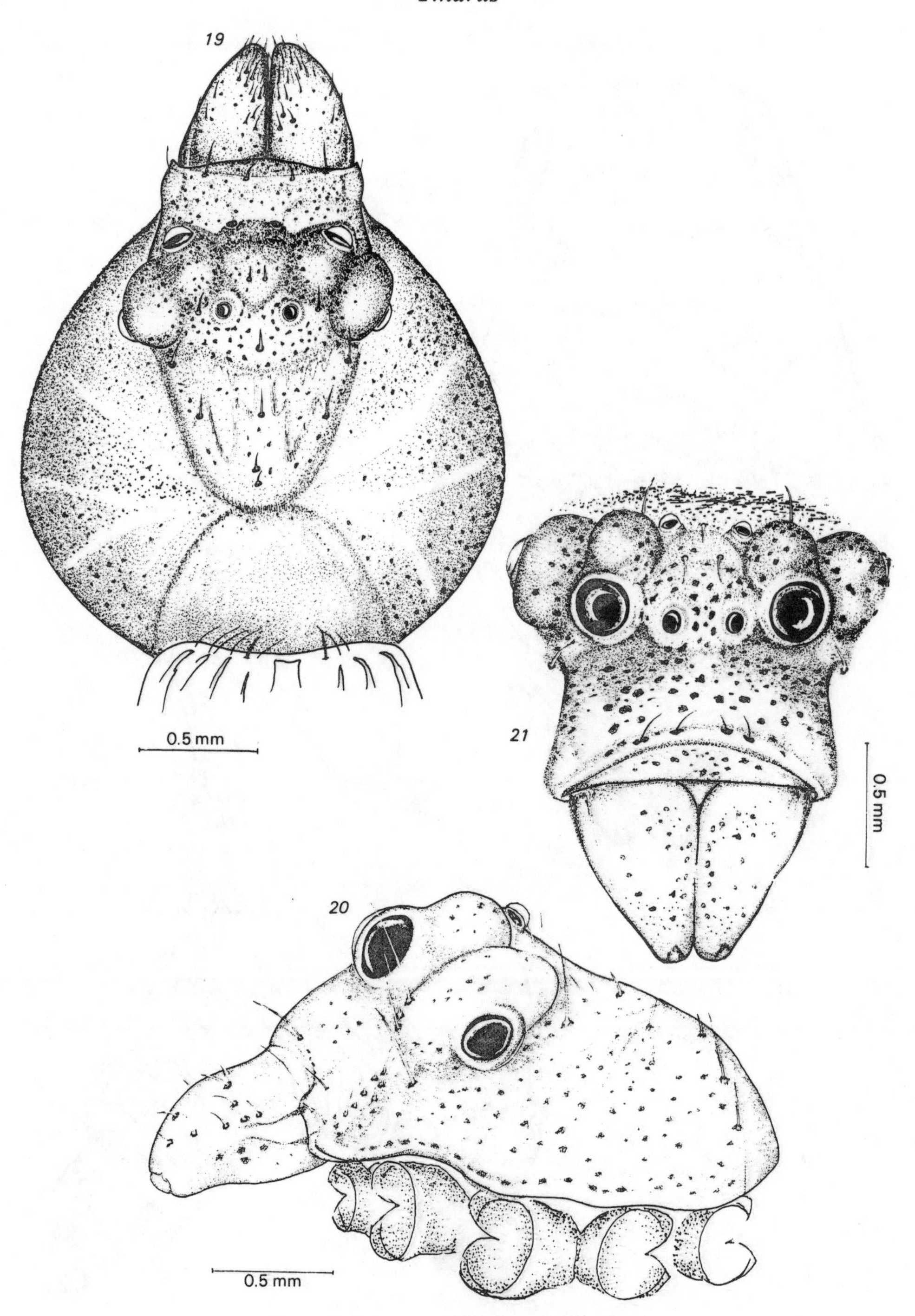

Figs. 19–21: *Tmarus* Simon, 1875; female
19. carapace, dorsal view; 20. prosoma, lateral view;
21. carapace, frontal view

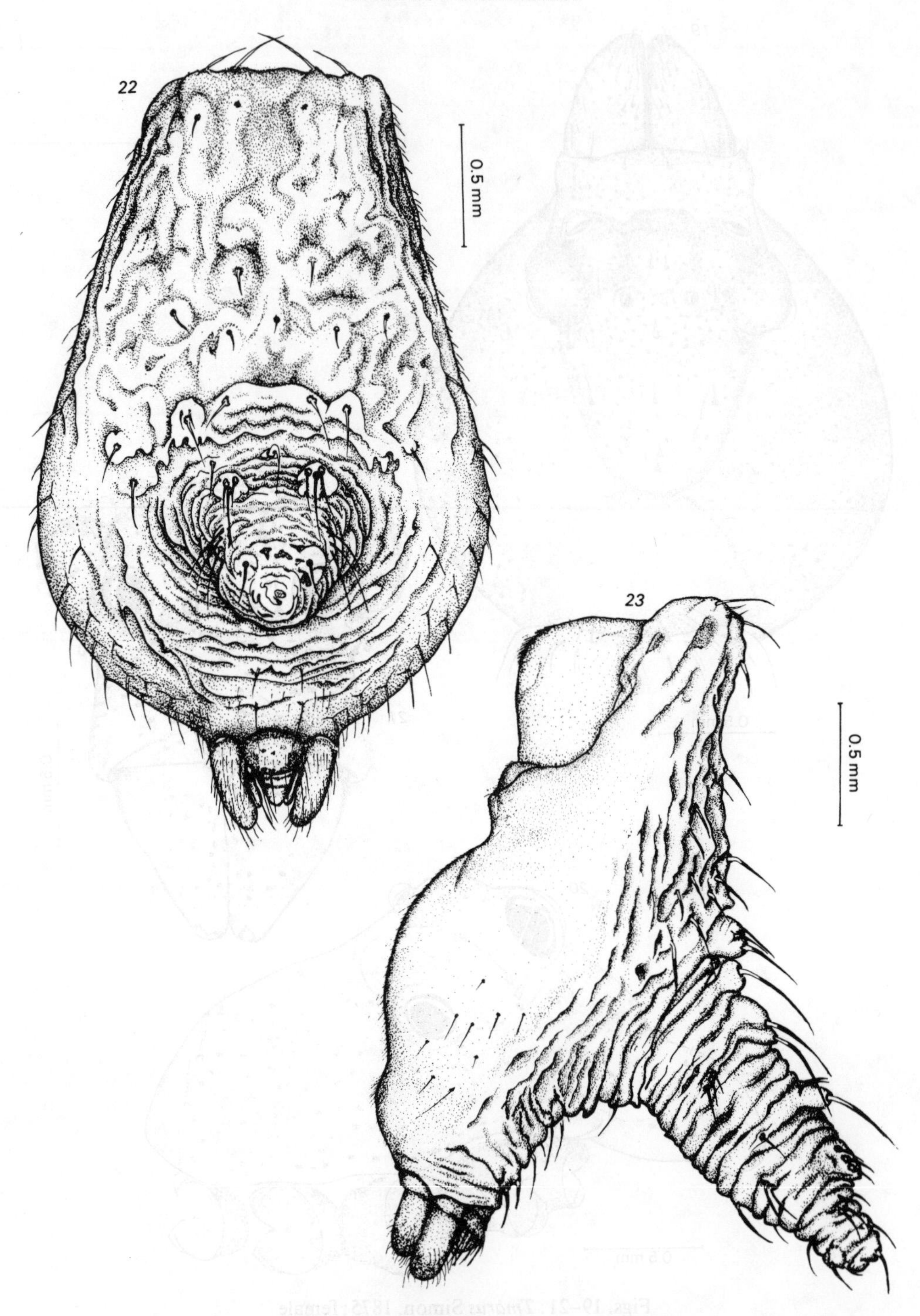

Figs. 22–23: *Tmarus* Simon, 1875; female, opisthosoma
22. dorsal surface; 23. lateral view

Tmarus piochardi (Simon, 1866)
Figs. 24–27

Thomisus piochardi Simon, 1866, *Annls Soc. ent. Fr.*, (4) 6 : 284.
Tmarus piochardi —. Simon, 1875, *Les Arachnides de France*, 2: 261; Roewer, 1954, *Katalog der Araneae*, 2 (1): 816; Bonnet, 1959, *Bibliographia Araneorum* 2 (5): 4645; Levy, 1973, *Israel J. Zool.*, 22: 116.
Tmarus piochardi var. *judaorum* Strand, 1915, *Arch. Naturgesch.*, 81 : 147 (misdetermination).

Length of male 5.0 mm, female 5.0–7.0 mm. Coloration grey with scattered dark, round patches and dots. Opisthosoma with a large wrinkled tubercle dorsally, and usually with a dark band, slightly broader at middle, along entire length on venter.
Male Palpus: Ventral tibial apophysis evenly thick, finger-like; stem slightly curved at middle with rounded top bent mesally (Fig. 24). Retrolateral tibial apophysis black,

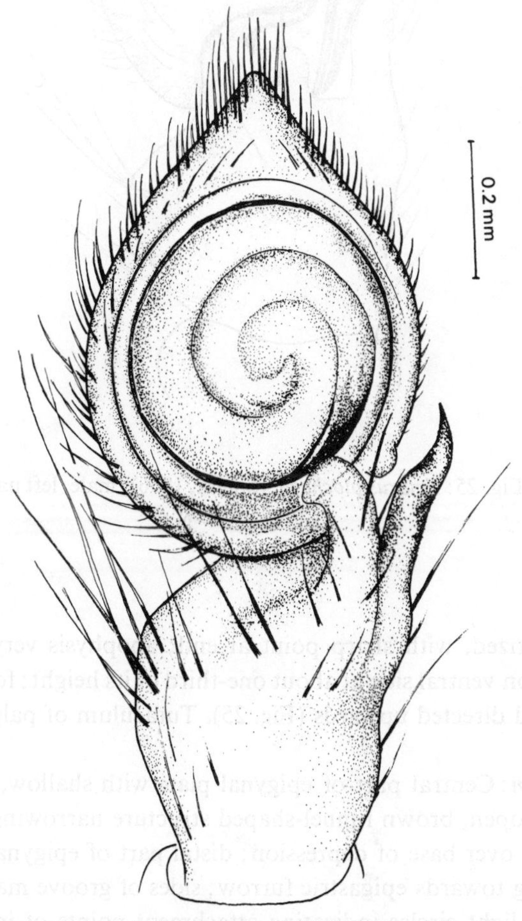

Fig. 24: *Tmarus piochardi* (Simon, 1866); male, left palpus, ventral view

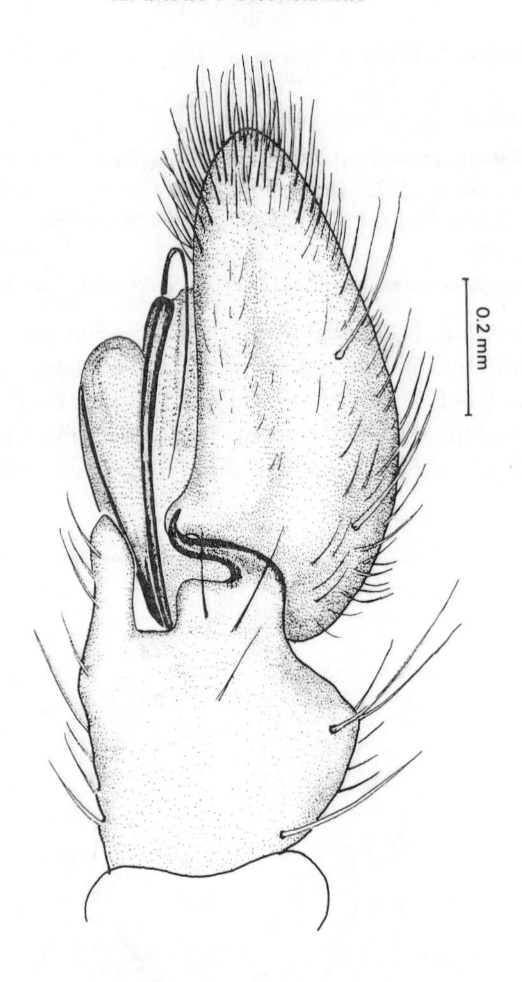

Fig. 25: *Tmarus piochardi* (Simon, 1866); male, left palpus, lateral view

markedly sclerotized, with sharp point at end; apophysis very broad at base and deeply notched on ventral side at about one-third of its height; following part of stem sharply bent and directed upwards (Fig. 25). Tutaculum of palpus short and broad (Fig. 24).

Female Epigynum: Central part of epigynal plate with shallow, rounded depression (Fig. 26); wide, open, brown funnel-shaped structure narrowing distally, projecting with a dark rim over base of depression; distal part of epigynal plate with shallow groove extending towards epigastric furrow; sides of groove marked with dark lines ending in small light circles indicating attachment points of internal spermathecal apodemes (Fig. 26; dorsal-inner view, Fig. 27).

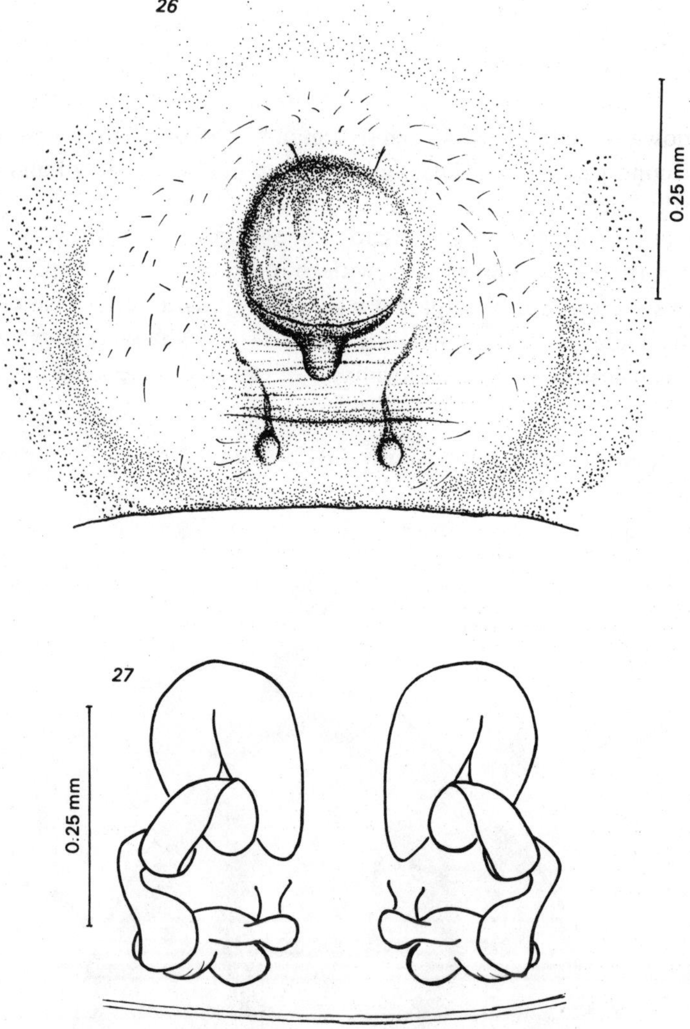

Figs. 26–27: *Tmarus piochardi* (Simon, 1866); female
26. epigynum; 27. spermathecae, inner view

Distribution : Southern Europe and northern Africa, Israel, Yemen.
Israel: Golan Heights, west of Quneitra (18), Mt. Tabor (2), Mt. Carmel (3) to Jerusalem (11).
Adults occur from April to June.

Tmarus yerohamus Levy, 1973
Figs. 28, 29

Tmarus yerohamus Levy, 1973, *Israel J. Zool.*, 22: 117.

Male unknown. Length of female 5.4–6.7 mm. Coloration light brown with numerous dark, round patches and dots. Opisthosoma grey with a dark longitudinal band on venter; dorsum with a large caudal tubercle.
Female Epigynum: Epigynal plate slightly cone-shaped, dark brown with markedly sclerotized parts; central part of plate with rounded depression, bordered distally by cup-shaped structure with transparent walls; pair of dark bulging bands rising from inside of cup-shaped structure, bending outwards and obliquely backwards, extending towards epigastric furrow and forming two elevated ridges (Fig. 28; dorsal-inner view, Fig. 29).

Distribution, Israel: Sand dunes south of Be'er Sheva' (15) April; Yeroḫam, Central Negev (17) June.

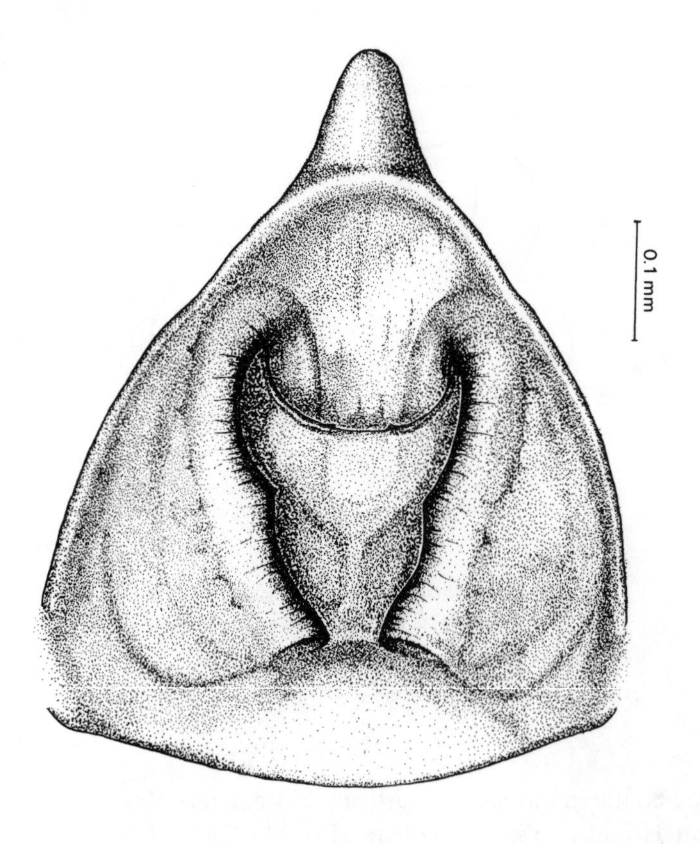

Fig. 28: *Tmarus yerohamus* Levy, 1973; female epigynum

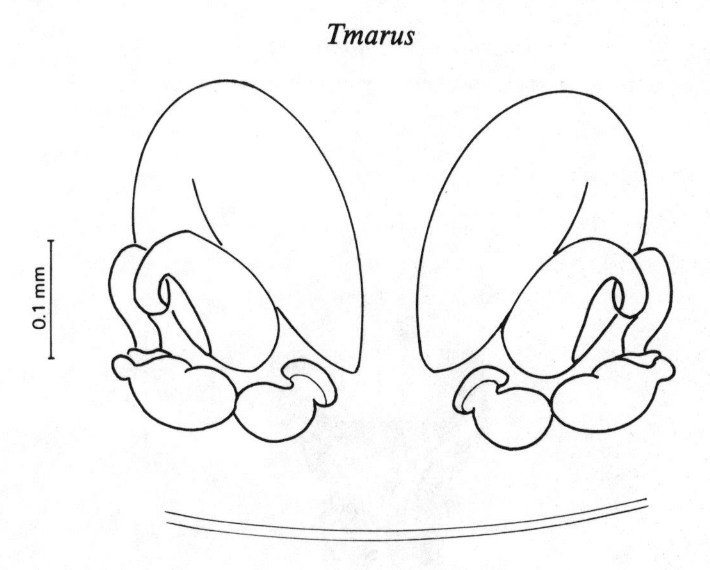

Fig. 29: *Tmarus yerohamus* Levy, 1973; female, spermathecae, inner view

Tmarus hazevensis Levy, 1973
Figs. 30, 31

Tmarus hazevensis Levy, 1973, *Israel J. Zool.*, 22: 119.

Male unknown. Length of female 6.0 mm. Coloration grey to yellowish-brown. Legs light yellow. Opisthosoma with scattered dark patches and a dark median band along venter; dorsum with a large caudal tubercle.
Female Epigynum: Basal part of epigynal plate with oval-shaped depression, slightly wider than long; at bottom of depression is situated a slightly elevated convex structure, distally pointed, with heavily sclerotized side walls; two dark, short, oval-shaped bulges obliquely set, protruding on both sides of pointed end of convex structure; distal pointed ends of bulges entering two adjacent sclerotic, protruding rings (Fig. 30; dorsal-inner view, Fig. 31).

Distribution, Israel: Ḫaẕeva, 'Arava Valley (14).
Adult females were found in April and October.

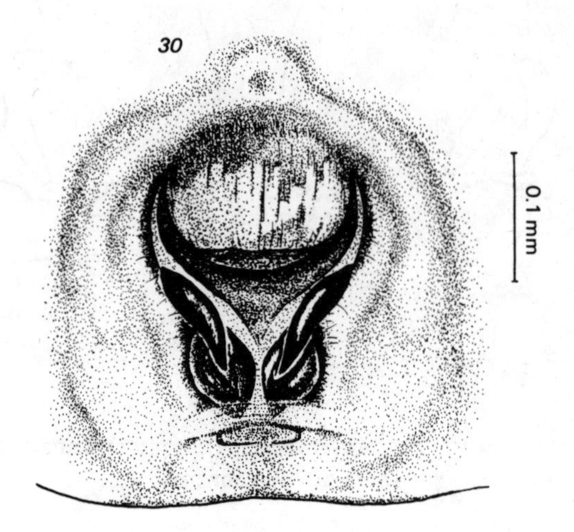

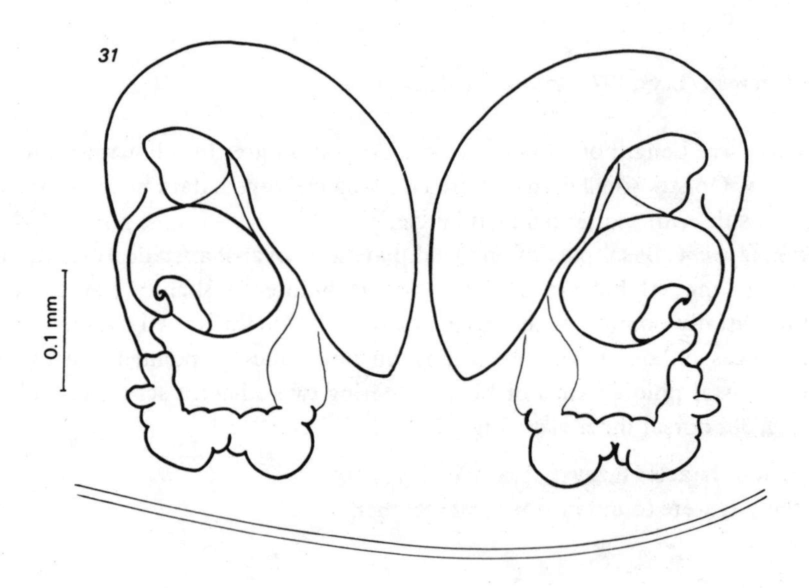

Figs. 30–31: *Tmarus hazevensis* Levy, 1973; female
30. epigynum; 31. spermathecae, inner view

Genus FIRMICUS Simon, 1895

Histoire Naturelle des Araignées, Paris, 1 (4): 1036

Figs. 32–35

Type Species: *Thomisus bragantinus* Capello, 1866.

Spiders with an elongated and extremely flat body. Carapace distinctly longer than wide, very flat and thin; frontal part of carapace very narrow, directed downwards and slightly backwards; tubercles of anterior-lateral eyes shallow and clearly larger than those of posterior-lateral eyes; anterior-lateral eyes larger than all other eyes;

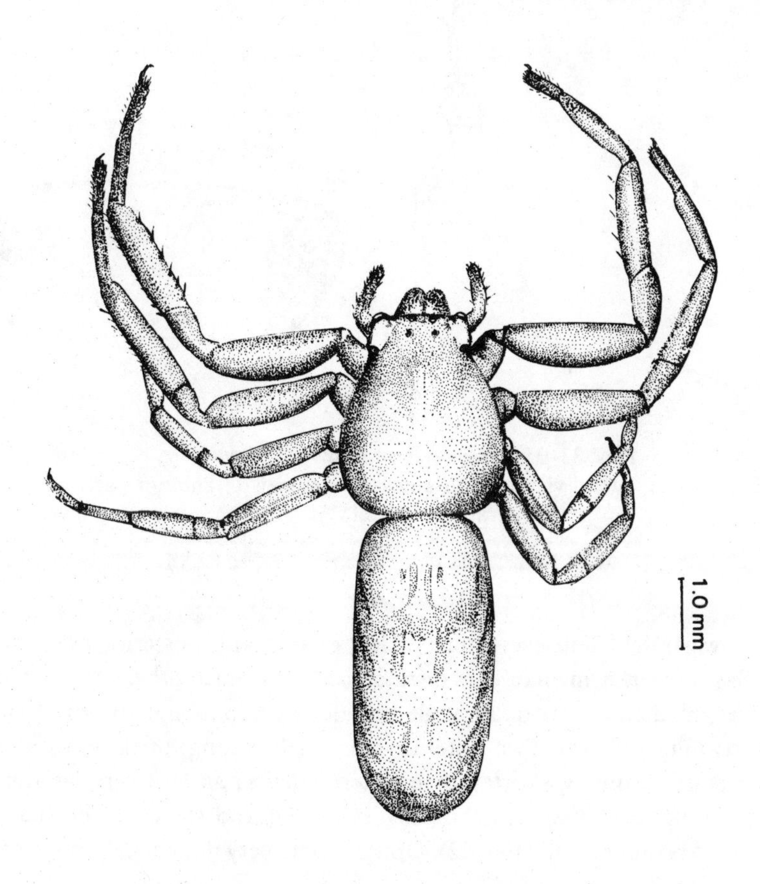

Fig. 32: *Firmicus* Simon, 1895; female, dorsal view of spider

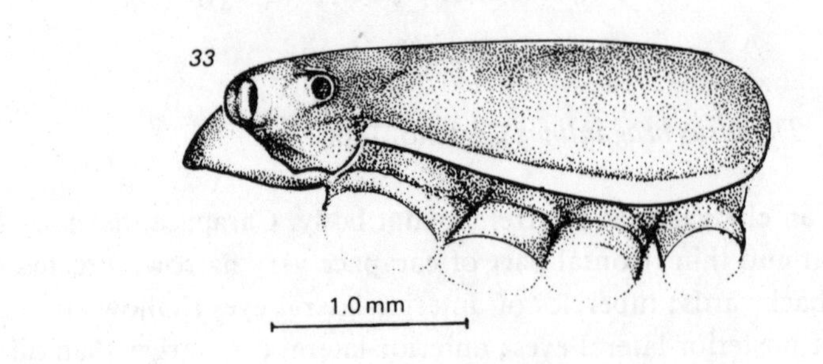

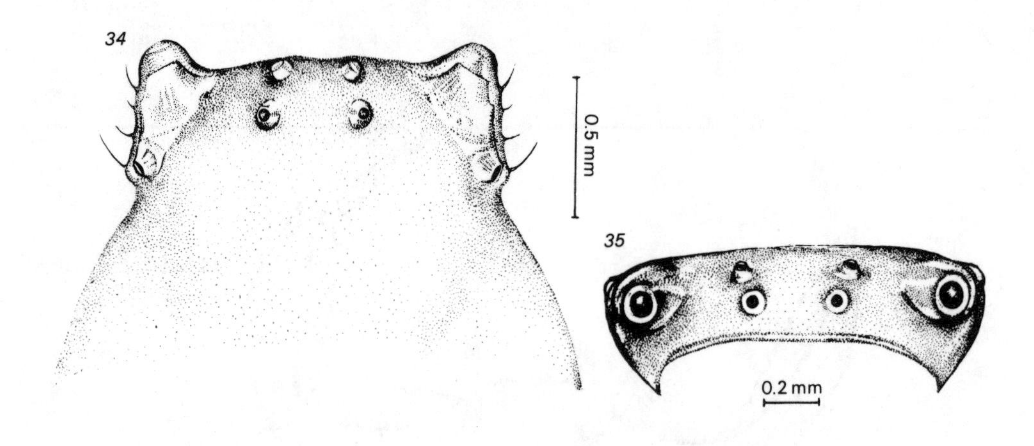

Figs. 33–35: *Firmicus* Simon, 1895; female
33. prosoma, lateral view; 34. carapace, dorsal view of anterior part;
35. carapace, frontal view

anterior eye row straight and eyes more or less equidistant; posterior eye row recurved (Fig. 34); posterior-median eyes distinctly closer to each other than to posterior-lateral eyes; median ocular quadrangle broader than long, narrower anteriorly than posteriorly (Figs. 33–35). Two anterior pairs of legs long, thick and spinous; two posterior pairs distinctly shorter than anterior pairs; all legs very laterigrade (Fig. 32); claws with denticles. Opisthosoma elongated and slender, sides parallel except at rounded posterior end (Fig. 32). Opisthosoma very flattened dorso-ventrally; spinnerets not visible from above.

Firmicus spiders are rare in Israel. They live under the bark of trees and under stones; their natural history is unknown.

About 20 species distributed mainly in Africa. One species in Israel.

Firmicus dewitzi Simon, 1899
Figs. 36–39

Firmicus dewitzi Simon, 1899, *Bull. Soc. ent. Fr.*, (13): 244; Roewer, 1954, *Katalog der Araneae*, 2 (1): 833; Bonnet, 1956, *Bibliographia Araneorum*, 2 (2): 1909; Levy, 1973, *Israel J. Zool.*, 22: 122.

Length of male 4.5–5.5 mm, female 5.1–6.8 mm. Coloration yellowish to almost white, sometimes opisthosoma greenish-black. Tubercles of eyes white in male. Opisthosoma with three pairs of longitudinal grooves on dorsum and a few small red dots close to distal end; venter with two medial rows of small depressions and a single, larger groove in front of the spinnerets.

Male Palpus: Relatively small. Ventral tibial apophysis low and broad, slightly inclined laterally and bent inwards apically (Figs. 36, 37). Intermediate and retrolateral tibial apophyses united at base. Intermediate apophysis directed laterally and bent inwards apically. Retrolateral apophysis large, bent laterally and pointed at end (Figs. 36, 37). Palpus deeply notched beneath tutaculum; posterior border of notch extending beyond retrolateral apophysis (Fig. 37).

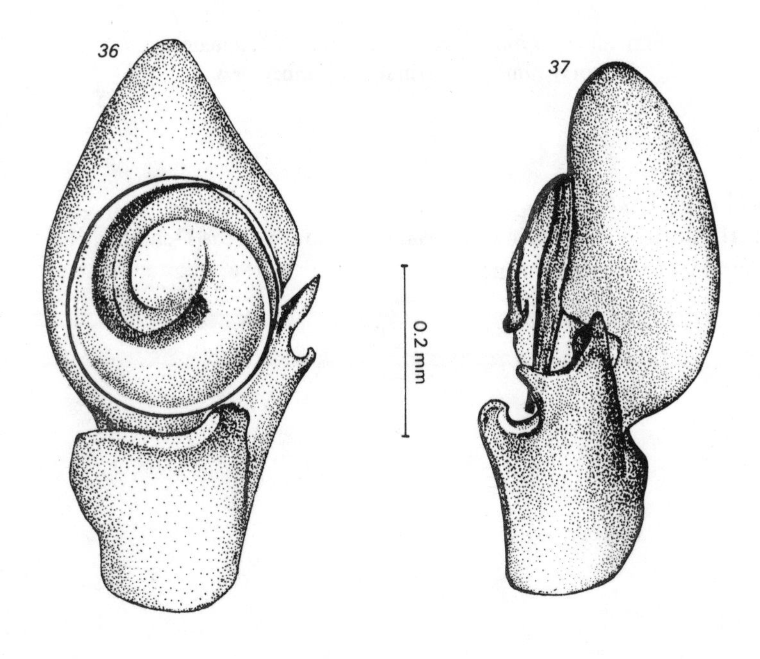

Figs. 36–37: *Firmicus dewitzi* Simon, 1899; male, left palpus
36. ventral view; 37. lateral view

Female Epigynum: Epigynal plate almost unpigmented; basal part with small transparent hood, opening distally; U-shaped sclerotized bands with an opening directed basally are positioned on both sides of hood's base; spermathecal bends visible through transparent epigynal plate, forming large semicircle centrally, near epigastric furrow (Fig. 38; dorsal-inner view, Fig. 39).

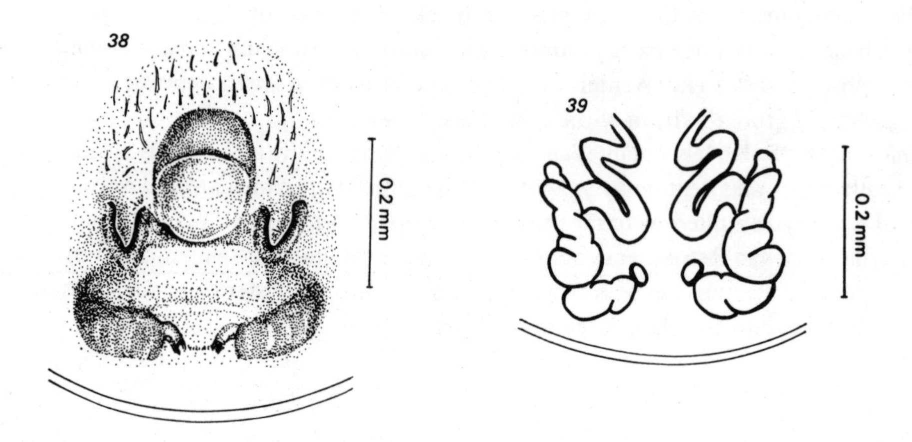

Figs. 38–39: *Firmicus dewitzi* Simon, 1899; female
38. epigynum; 39. spermathecae, inner view

Distribution: Egypt, Israel.
Israel: Along the Jordan Rift, from Ginnosar (7) to south of the Dead Sea (13).
Adults are found from March to June.

Genus THOMISUS Walckenaer, 1805

Tableau des Aranéides, Paris, p. 28

Figs. 40–42

Type Species: *Thomisus onustus* Walckenaer, 1805.

Spiders with a short, stout body. Carapace rather high, convex and approximately as long as wide; clypeus vertical (Fig. 41). Lateral eyes on each side borne on single common tubercle; tubercles of lateral eyes large, and markedly protuberant at anterior-lateral corners of carapace; lateral eyes on side of tubercles, not on top (Figs. 40, 41). All eyes rather small; anterior row of eyes recurved; anterior-median eyes usually closer to each other than to anterior-lateral eyes; posterior row of eyes straight or slightly recurved; posterior-median eyes farther from each other than from posterior-lateral eyes; eyes of posterior row equidistant; median ocular quadrangle broader posteriorly than anteriorly (Figs. 40, 41). Legs thick and long; two anterior pairs distinctly longer than two posterior pairs; claws with denticles. Opisthosoma triangular, narrow anteriorly, almost truncated and very broad posteriorly; dorsally flat or slightly convex, ventrally rounded (Fig. 42). Hairs, bristles and occasionally small spines on body.

Thomisus spiders usually live in vegetation, mainly inside flower corollas. One species is known for its ability to change colours to match the substratum (see above, p. 13). They catch their prey lurking on the flowers with their legs spread widely.

More than 100 species distributed in greater parts of the Old World. Three species in Israel.

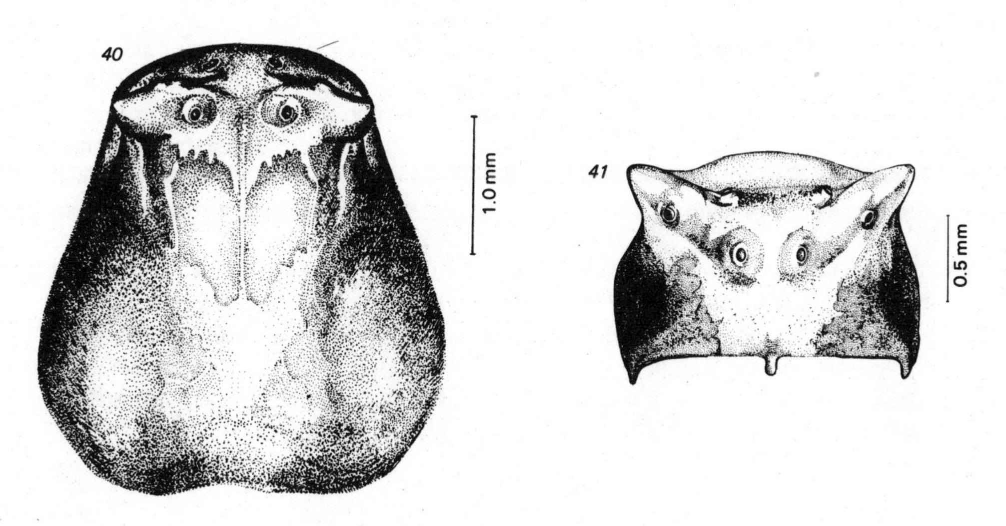

Figs. 40–41: *Thomisus* Walckenaer, 1805; female
40. carapace, dorsal view; 41. carapace, frontal view

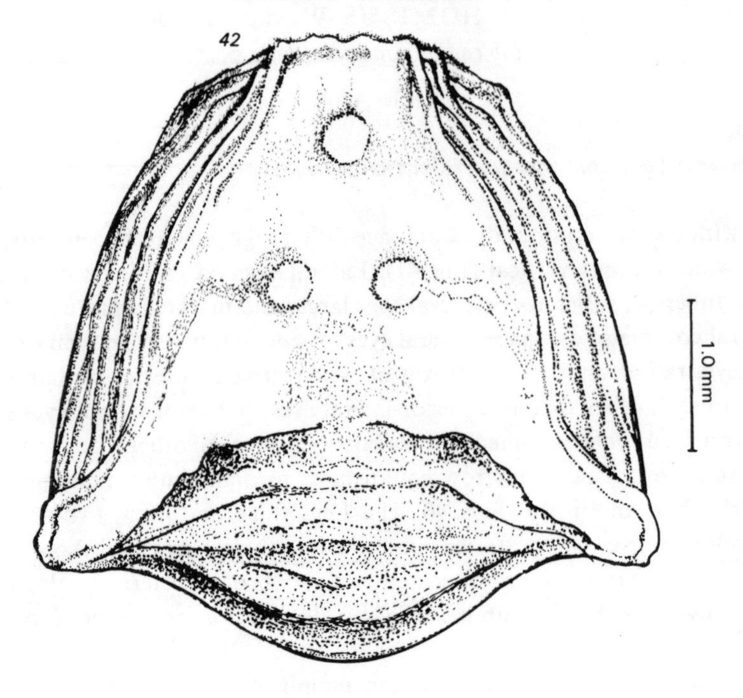

Fig. 42: *Thomisus* Walckenaer, 1805; female, opisthosoma, dorsal surface

Key to the Species of Thomisus in Israel and Sinai

1. Patella of male palpus (not the tibia) with two large, distinct spines on lateral side (Figs. 51, 52). Thick, sclerotic embolus of palp rises at distal (apical) part of tegulum (Fig. 51). Epigynum of female with a distinct, raised, centrally constricted, median septum (Fig. 53). **T. bidentatus** Kulczynski
- Patella of palpus without conspicuous spines. Thick sclerotic embolus rises at proximal (basal) part of tegulum (Figs. 43, 47). Epigynum smooth, without a median septum (Figs. 45, 49). 2
2. Tegulum of male palpus on latero-basal margin with a large, pointed apophysis projecting towards tibia of palpus (Figs. 47, 48). Epigynal plate of female with a large, oval-shaped swelling located between the dark attachment marking of the spermathecal apodemes and the epigastric furrow (Fig. 49). **T. citrinellus** Simon
- Tegulum of male palpus without an apophysis. Epigynal plate of female shield-shaped, markedly convex and gradually sloping towards epigastric furrow (Fig. 45). **T. onustus** Walckenaer

Thomisus onustus Walckenaer, 1805
Figs. 43–46

Thomisus onustus Walckenaer, 1805, *Tableau des Aranéides*, Paris, p. 32. Full synonymy in: Roewer, 1954, *Katalog der Araneae*, 2 (1): 856 and Bonnet, 1959, *Bibliographia Araneorum*, 2 (5): 4585; Levy, 1973, *Israel J. Zool.*, 22: 124.

Thomisus peronii Audouin, 1827. O. P.-Cambridge, 1872, *Proc. zool. Soc. Lond.*, p. 308. Bonnet, 1959, *Bibliographia Araneorum*, 2 (5): 4589 (misdetermination).

Thomisus abbreviatus Walckenaer, 1837. O. P.-Cambridge, 1872, *Proc. zool. Soc. Lond.*, p. 308.

Pistius truncatus (Pallas, 1772). Pavesi, 1895, *Boll. Musei Zool. Anat. Comp. R. Univ. Torino*, 10: 8 (as *Misumena truncata*; misdetermination).

Misumena vatia (Clerck, 1757). Pavesi, 1895, *Boll. Musei Zool. Anat. Comp. R. Univ. Torino*, 10: 8 (misdetermination).

Length of male 2.0–3.6 mm, female 7.0–9.8 mm. Coloration white or yellow; occasional specimens with pink patches all over. Prosoma and legs of males often dark brown.

Male Palpus: Ventral and retrolateral tibial apophyses with series of dark, small tubercles; long bristles on top of tubercles, along outer sides of apophyses and basal

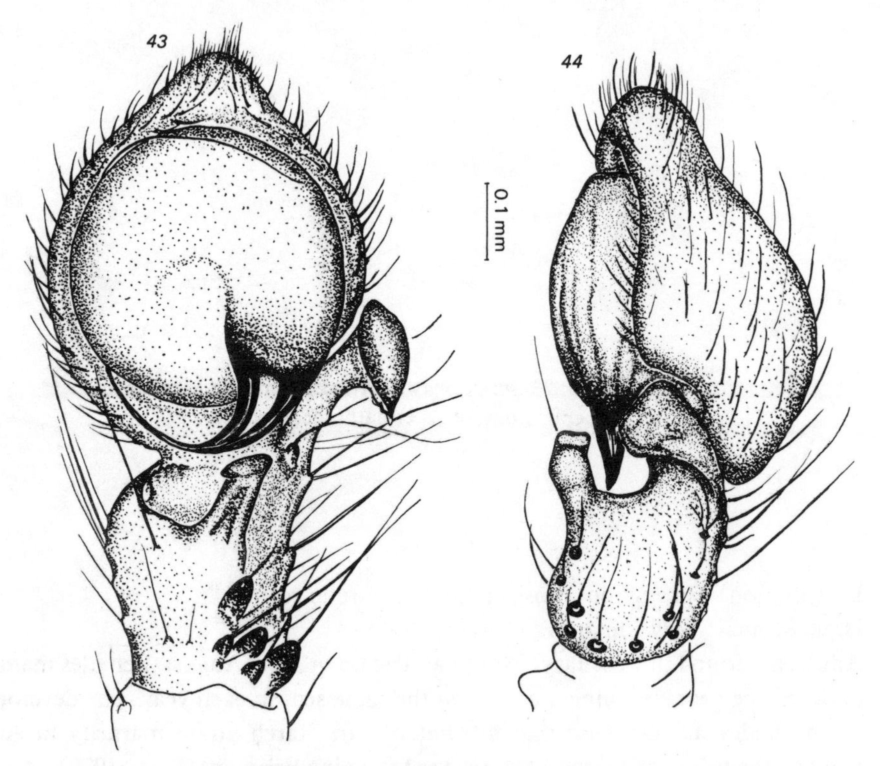

Figs. 43–44: *Thomisus onustus* Walckenaer, 1805; male, left palpus
43. ventral view; 44. lateral view

part of tibia. Ventral apophysis slightly oblique, bent laterally; apex slightly swollen and bent mesally. Retrolateral apophysis with slight lateral bend, slender at middle, with sharp ventral bend in apical part (Figs. 43, 44). Palpus beneath tutaculum deeply notched; notch extending beyond retrolateral apophysis (Fig. 44).

Female Epigynum: Epigynal plate shield-shaped, markedly convex, occasionally dark and heavily sclerotized; spermathecal orifices on both sides of basal part of epigynal plate; orifices large, opening obliquely upwards and outwards, with dark, narrow duct at middle of each opening; attachment points of spermathecal apodemes marked by two dark, rather conspicuous circles on distal part of epigynal plate (Fig. 45; dorsal-inner view, Fig. 46 : circles in latter figure represent balls of extremely convoluted ducts, not shown in drawing).

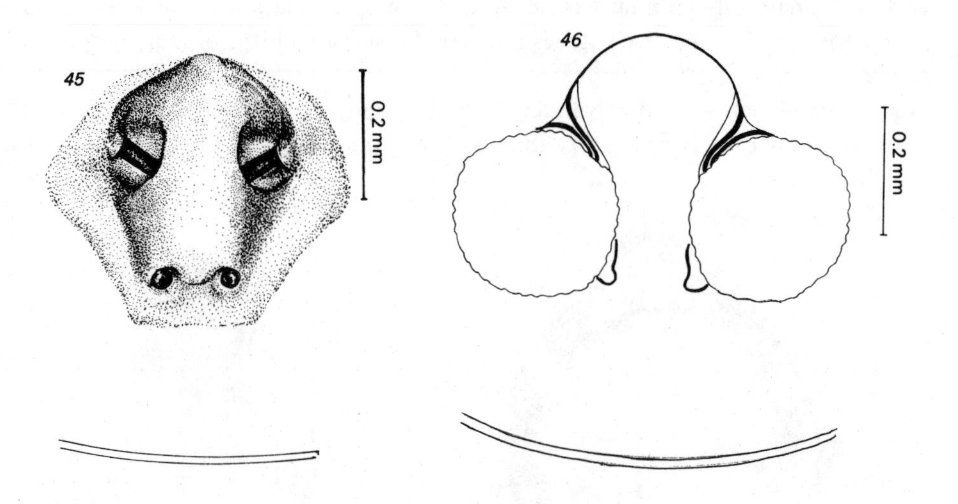

Figs. 45–46: *Thomisus onustus* Walckenaer, 1805; female
45. epigynum; 46. spermathecae, inner view

Distribution: Throughout most of the Palaearctic region.

Israel & Sinai: Very common throughout.

Adults are found in February–March and again in July–August. Females maintain a cycle of one year, reaching maturity in the same season each year. The development of the males differs: those that are hatched in March attain maturity in August, whereas the brood of August matures the following February (Levy, 1970).

Thomisus citrinellus Simon, 1875
Figs. 47–50

Thomisus spinifer O. P.-Cambridge, 1872, *Proc. zool. Soc. Lond.*, p. 308 (preoccupied).
Thomisus citrinellus Simon, 1875, *Les Arachnides de France*, 2: 253; Roewer, 1954, *Katalog der Araneae*, 2 (1): 857; Bonnet, 1959, *Bibliographia Araneorum*, 2 (5): 4578; Levy, 1973, *Israel J. Zool.* 22: 127.

Length of male 2.1–2.6 mm, female 5.0–7.3 mm. Coloration whitish-yellow. Females, occasionally, with a round, black spot on each conical protuberance on posterior corners of opisthosoma. Prosoma and legs rather spinose, mainly in males.
Male Palpus: Tibia with ventral apophysis only; on lateral side two slightly curved, strong spines with dark, pointed tips; ventral apophysis in form of round, stemless, hollow swelling (Figs. 47, 48). Tegulum on latero-basal margins with large, downwards inclined tegular apophysis extending with pointed tip to almost half the length of tibia (Fig. 48); tutacular area of palpus smooth and intact (Fig. 48).

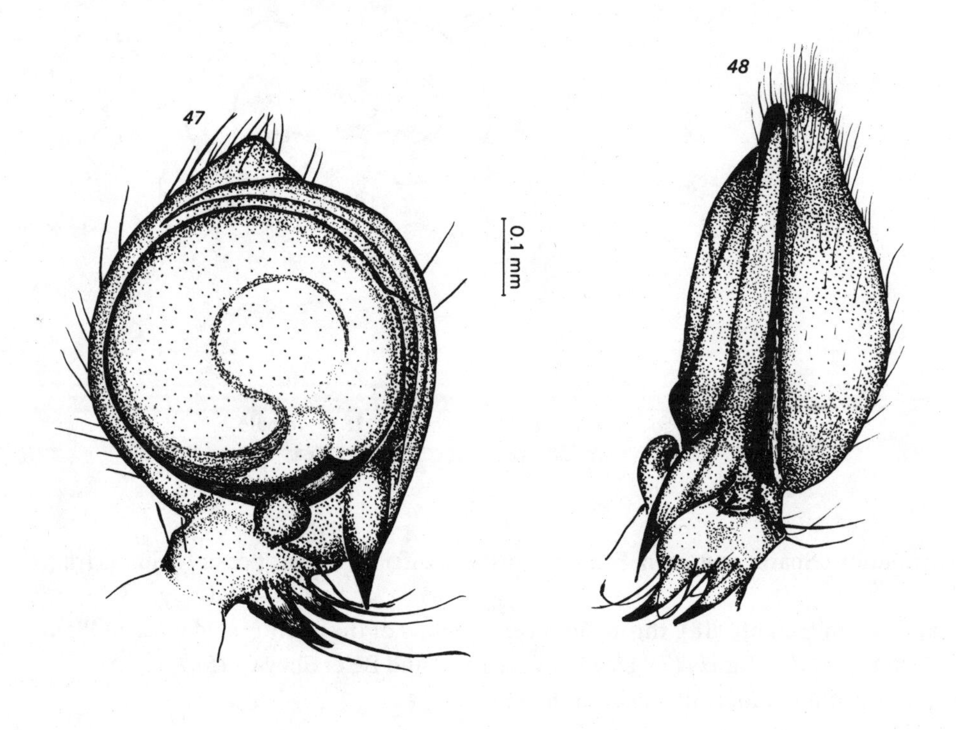

Figs. 47–48: *Thomisus citrinellus* Simon, 1875; male, left palpus
47. ventral view; 48. lateral view

Female Epigynum: Epigynal plate almost unpigmented; distal and central parts of plate slightly convex, forming a large, oval swelling; two dark, round, adjacent spermathecal orifices are located medially, above swelling; dark ducts, directed obliquely outwards, extend from each opening (Fig. 49; dorsal-inner view, Fig. 50).

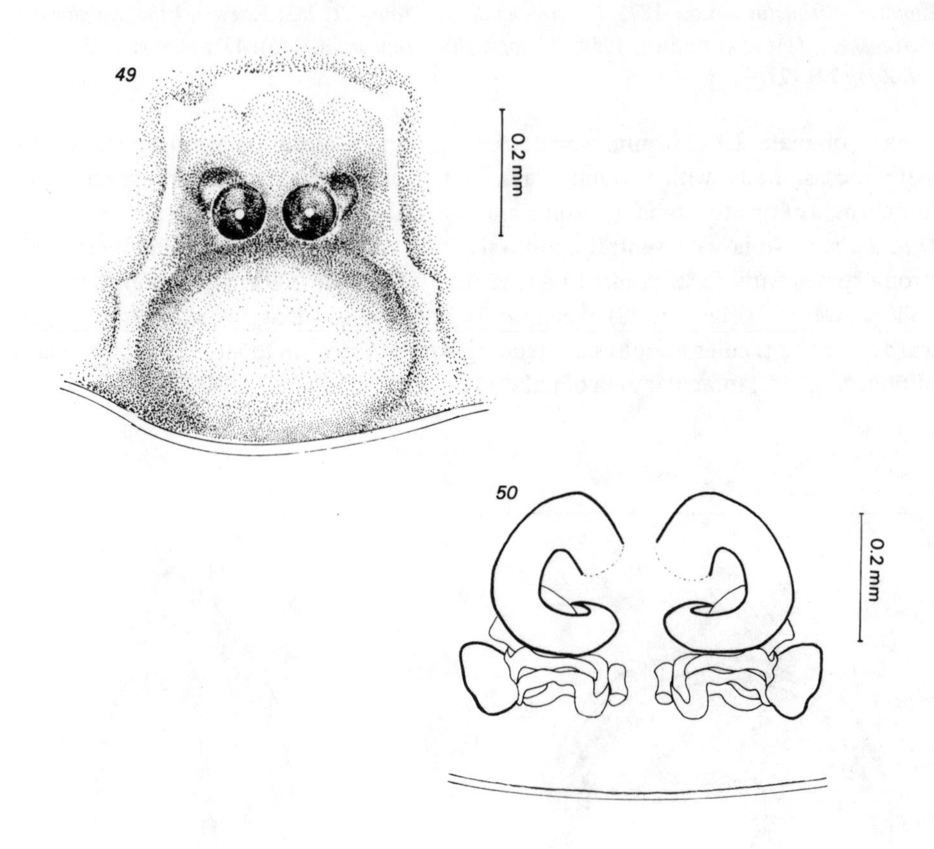

Figs. 49–50: *Thomisus citrinellus* Simon, 1875; female
49. epigynum; 50. spermathecae, inner view

Distribution: Spain, southern France, northern Africa, Israel, possibly Sinai, Iraq, Yemen.
Israel: Golan Heights (18), throughout central parts of the country and Coastal Plain (8, 9) to the Jordan Valley (7), Dead Sea area (13) and Be'er Sheva' (15).
Adults are found almost throughout the year.

Thomisus bidentatus Kulczyński, 1901
Figs. 51–54

Thomisus bidentatus Kulczyński, 1901, *Rozpr. Akad. Krakow.*, 41 : 3, 33; Roewer, 1954, *Katalog der Araneae*, 2 (1): 858; Bonnet, 1959, *Bibliographia Araneorum*, 2 (5): 4577; Levy, 1973, *Israel J. Zool.*, 22: 129.

Length of male 2.7–2.9 mm, female 4.4–4.7 mm. Coloration of prosoma intense green with white eye tubercles; opisthosoma green with yellowish-brown or red markings on back; legs green to light brown. Prosoma and legs spinose, mainly in males.

Male Palpus: Patella with two large, sharply pointed spines on lateral side, a few smaller spines on other parts of patella. Tibia with ventral and retrolateral apophyses; ventral apophysis small, hook-shaped, almost transparent; retrolateral apophysis large, broad, projecting slightly near apex (Figs. 51, 52). Tutacular area rather retracted beyond retrolateral apophysis (Fig. 52).

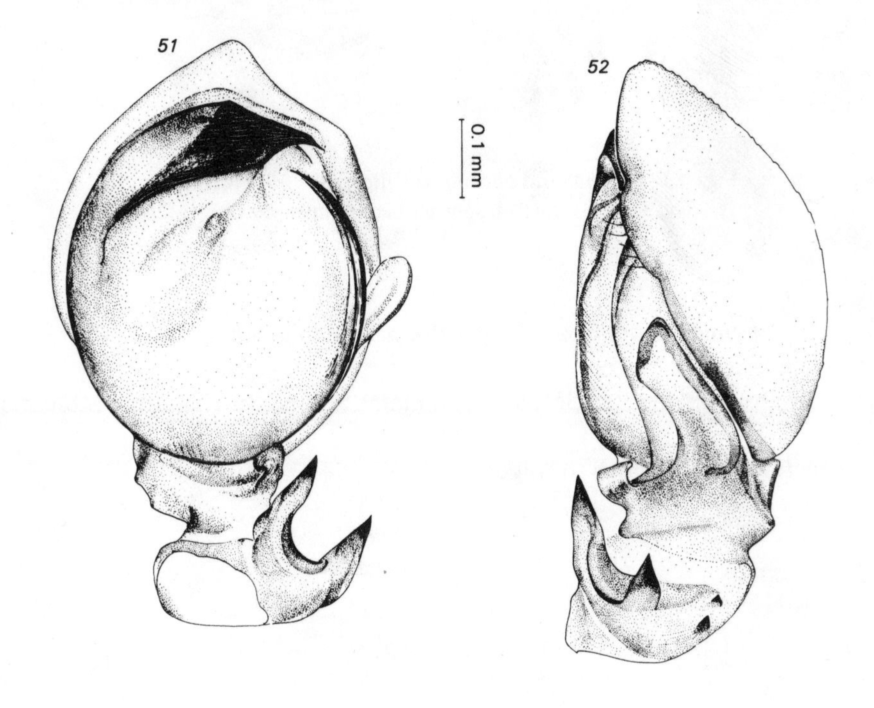

Figs. 51–52: *Thomisus bidentatus* Kulczyński, 1901; male, left palpus
51. ventral view; 52. lateral view

Female Epigynum: Distal part of epigynal plate with a broad, longitudinal depression surrounding a raised, rounded white area; basal part of raised area constricted, forming a median septum between two deep depressions containing brown cone-shaped openings of the spermathecae (Fig. 53; dorsal-inner view, Fig. 54).

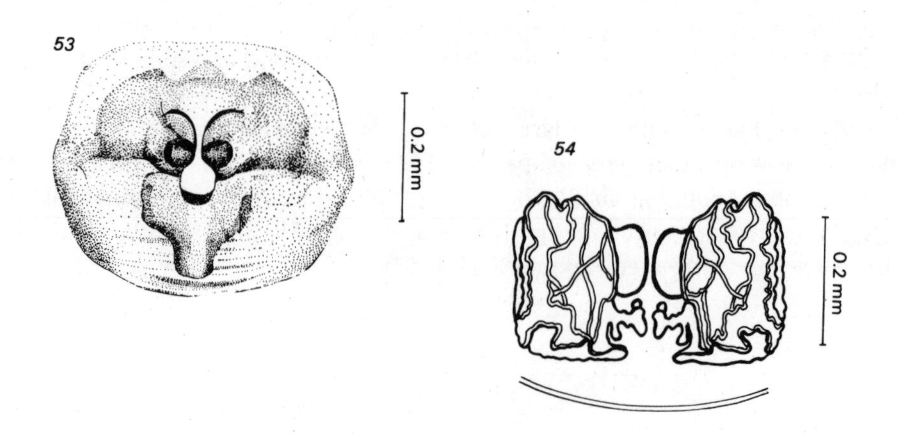

Figs. 53–54: *Thomisus bidentatus* Kulczyński, 1901; female
53. epigynum; 54. spermathecae, inner view

Distribution: Guinea, Ivory Coast, Upper Volta, Ethiopia, Sinai, Israel.
Israel: Wadi Qelt (12).
Sinai: Nuweiba near the Red Sea, and mountains around St. Katharina Monastery (22).
Adults are found from March to August.

Genus RUNCINIA Simon, 1875

Les Arachnides de France, 2: 254

Figs. 55, 56

Type Species: *Thomisus lateralis* C. L. Koch, 1838.

Spiders with thick, oval body. Carapace as long as wide or slightly longer than wide, rather low and partly flattened dorsally (Fig. 55); clypeus sharply truncated; upper fore-corners conically protuberant laterally (Figs. 55, 56). Anterior row of eyes borne on frontal slope while posterior eyes located on dorsal side of carapace; lateral eyes of each side borne on single, low common tubercle protruding laterally (Fig. 55); lateral eyes borne on side of tubercle, not on top, larger than median eyes; both

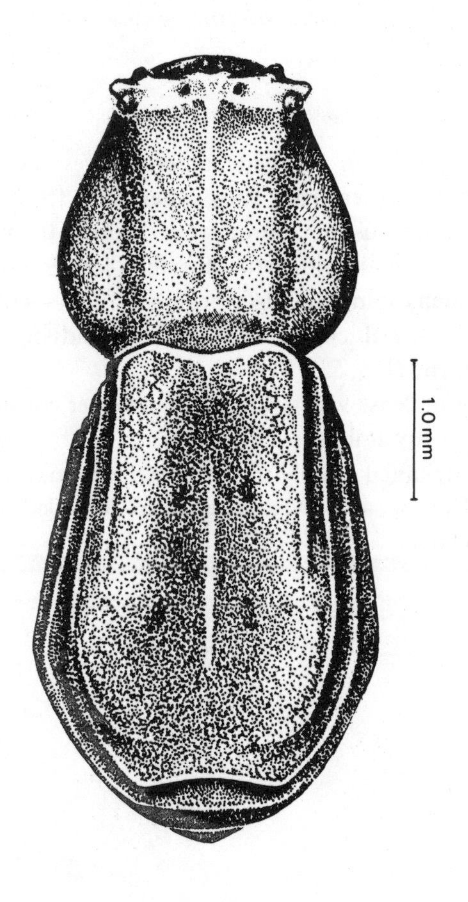

Fig. 55: *Runcinia* Simon, 1875; female, dorsal view of spider

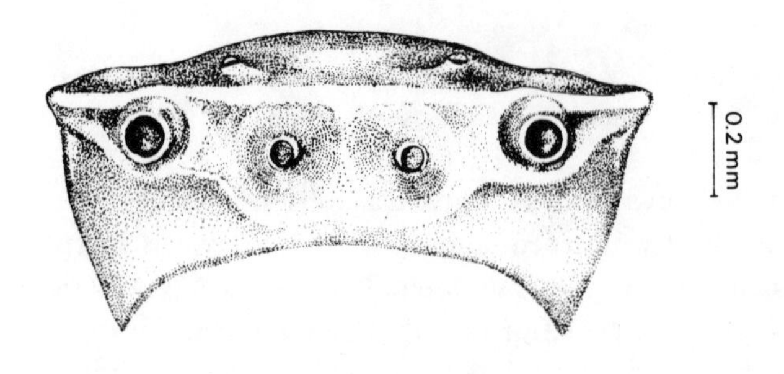

Fig. 56: *Runcinia* Simon, 1875; female, carapace, frontal view

eye-rows slightly recurved or straight; anterior eyes more or less equidistant; posterior-median eyes farther from each other than from posterior-lateral eyes; median ocular quadrangle distinctly broader posteriorly than anteriorly. Legs rather long and not slender; two anterior pairs distinctly longer than two posterior pairs; anterior pairs with many spines on distal segments; claws with a few large denticles. Opisthosoma oval, rather thick, dorsally flattened and slightly wider on posterior third and rounded at end (Fig. 55).

Runcinia spiders live in dense vegetation, also on flower corollas. Their movement is slow and prey is caught by ambush.

About 25 species distributed mainly in Africa. One species occurs in the Mediterranean and southern Europe, and several species are recorded from South and Central Asia. One species in Israel.

Runcinia lateralis (C. L. Koch, 1838)
Figs. 57–60

Thomisus lateralis C. L. Koch, 1838, *Die Arachniden*, Nürnberg, IV, p. 43.
Runcinia lateralis —. Simon, 1875, *Les Arachnides de France*, 2: 255; Roewer, 1954, *Katalog der Araneae*, 2 (1): 853 (as *R. cerina*); Bonnet, 1958, *Bibliographia Araneorum*, 2 (4): 3884; Levy, 1973, *Israel J. Zool.*, 22: 132.

Length of male 3.3–4.6 mm, female 5.4–7.3 mm. Coloration white or yellow with dark bands along sides of carapace and opisthosoma. Females usually with deep marginal folds around opisthosoma, and sometimes show a colourful pattern on dorsal surface. The young are able to change colour from white to yellow and vice versa, but only following a moult.
Male Palpus: Tibia with ventral and retrolateral apophyses. Ventral apophysis very short, in form of small hook, almost unnoticeable from below. Retrolateral apophysis cone-shaped, rather large, extending more than half the length of tegulum (Figs. 57, 58).

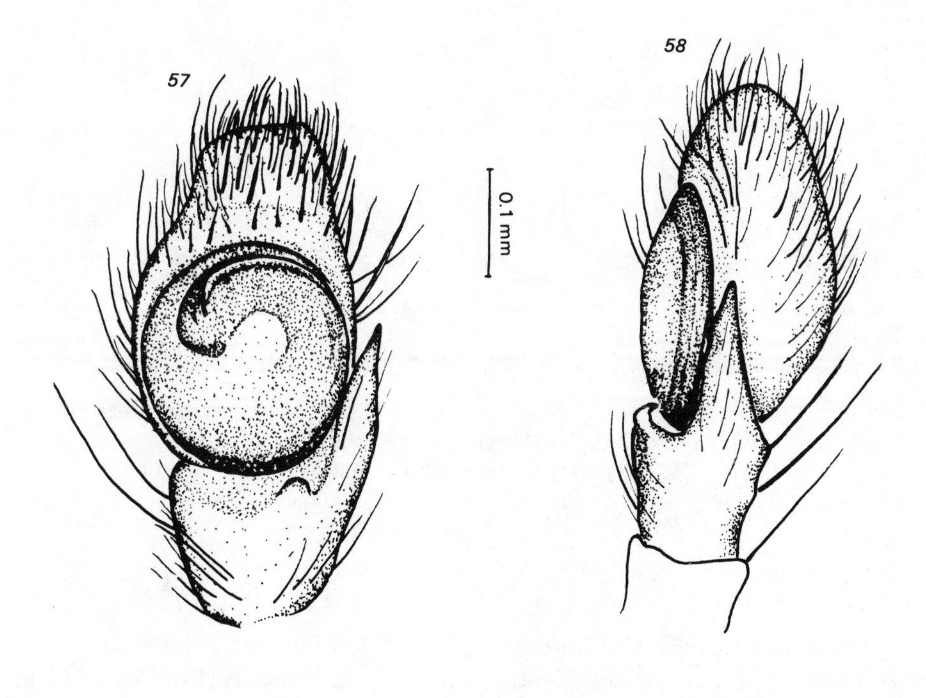

Figs. 57–58: *Runcinia lateralis* (C. L. Koch, 1838); male, left palpus
57. ventral view; 58. lateral view

Female Epigynum: Basal part of epigynal plate with convex, hood-shaped, transparent structure with opening directed towards epigastric furrow; pair of dark ridges surrounding distal margins of hood, forming two loops inside hood; distal part of epigynal plate with two very dark, sclerotic rings, marking the attachment points of the internal spermathecal apodemes (Fig. 59; dorsal-inner view, Fig. 60).

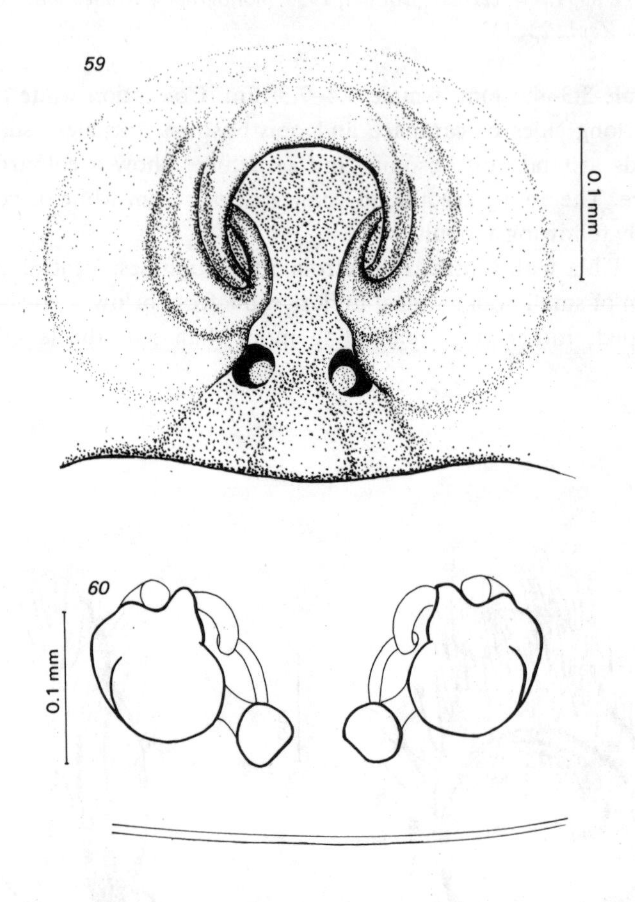

Figs. 59–60: *Runcinia lateralis* (C. L. Koch, 1838); female
59. epigynum; 60. spermathecae, inner view

Distribution: Southern Europe, northern Africa, Middle East to Central Asia.
Israel: Common throughout the country from the Golan Heights (18), Coastal Plain (8, 9), Judean Hills (11) to the Jordan Rift (7, 13).
Adult males occur from March to June, mainly from May to June; mature females are found from May to October, mainly from May to July.

Genus HERIAEUS Simon, 1875

Les Arachnides de France, 2: 203

Figs. 61, 62

Type Species: *Thomisus hirtus* Latreille, 1819.

Very hairy spiders, usually with stout body. Carapace rather high and slightly convex, truncated in front, and as long as wide or slightly wider than long (Figs. 61, 62). Eyes borne on distinctly separated tubercles; tubercles of lateral eyes large and protruding more than those of median eyes; anterior row of eyes more recurved than posterior row; anterior-median eyes usually farther from each other than from anterior-lateral eyes; posterior-median eyes usually closer to each other than to posterior-lateral eyes, or eyes equidistant; median ocular quadrangle narrower posteriorly than anteriorly, its lateral sides longer than posterior side (Figs. 61, 62). Legs rather long, thick, spinose and with numerous bristles; two anterior pairs distinctly longer than two posterior pairs; claws with few denticles. Opisthosoma oval or rounded. Body and legs covered all over with long, rather thick bristles; bristles occasionally borne on small tubercles.

Heriaeus spiders live in dense vegetation. Their body with its heavy pilosity strongly resembles certain seeds and they are scarcely detectable when not in motion. They move slowly and catch their prey by ambush. About 20 species distributed in Africa to Central Asia; some occur in Europe. Two species in Israel.

Key to the Species of Heriaeus in Israel

1. Tibia of male palpus with three distinct apophyses (Fig. 63); tip of thick, sclerotic embolus hardly visible (Figs. 63, 64). Epigynum of female with a convex, transparent, hood-shaped structure (Fig. 65); spermathecae accompanied by large, longitudinal folds (Fig. 66, require dissection. In northern and central parts of the country. **H. setiger** (O. P.-Cambridge)

– Tibia of male palpus with only two distinct apophyses, the third, the intermediate apophysis, being markedly reduced (Fig. 67); distinct tip of embolus is strongly recurved, forming a wide loop (Figs. 67, 68). Epigynum of female with a flat, sclerotic, hood-shaped structure (Fig. 69); spermathecae form two complete coils (Fig. 70, require dissection). In eastern and southern parts of the country. **H. buffoni** (Audouin)

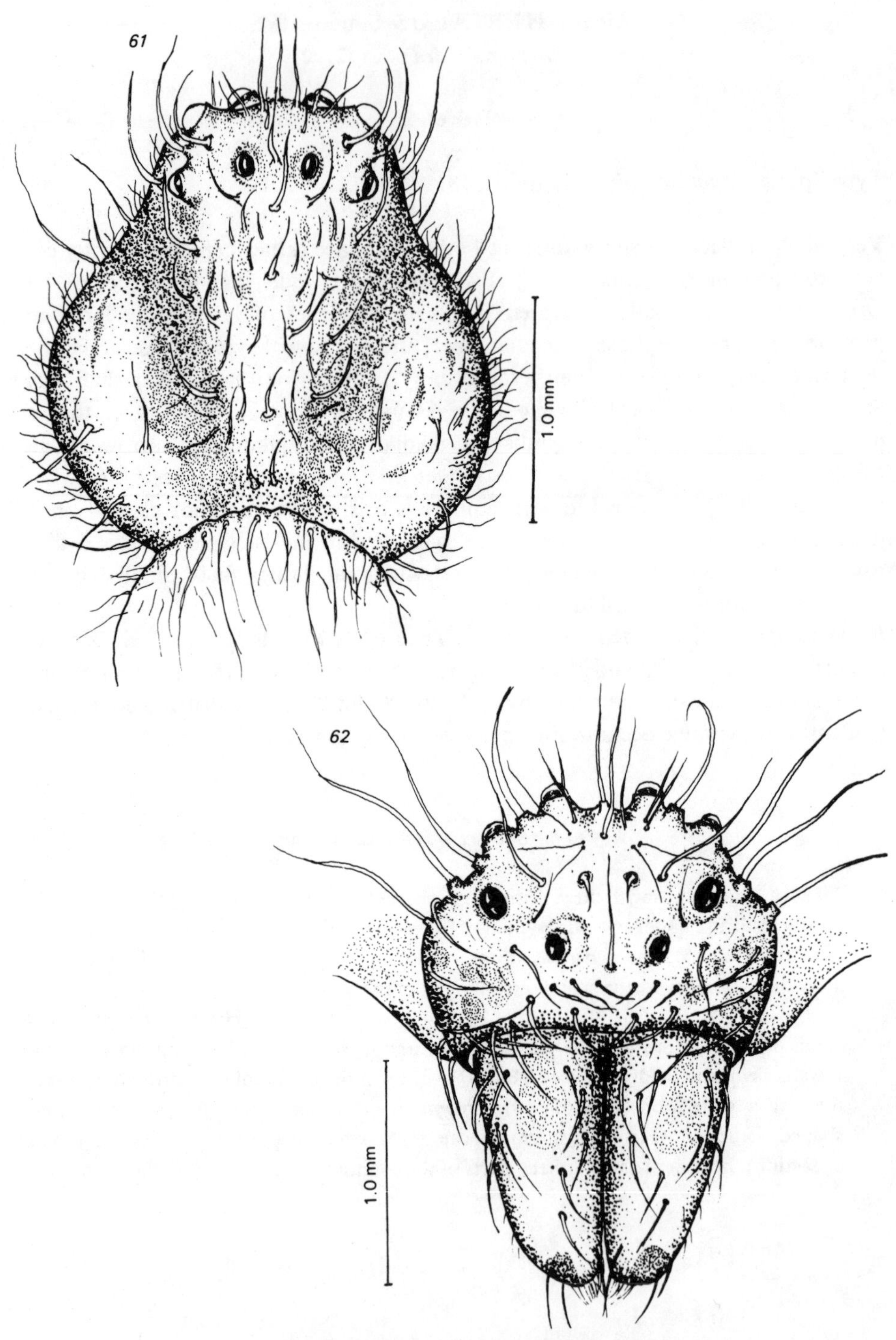

Figs. 61–62: *Heriaeus* Simon, 1875; male, carapace
61. dorsal view; 62. frontal view

Heriaeus setiger (O. P.-Cambridge, 1872)
Figs. 63–66

Thomisus setiger O. P.-Cambridge, 1872, *Proc. zool. Soc. Lond.*, p. 307.
Heriaeus setiger —. Simon, 1875, *Les Arachnides de France*, 2: 208; Roewer, 1954, *Katalog der Araneae*, 2 (1): 866; Bonnet, 1957, *Bibliographia Araneorum*, 2 (3): 2165; Levy, 1973, *Israel J. Zool.*, 22: 136.

Length of male 3.9–5.6 mm, female 5.5–9.2 mm. Coloration yellow, green or brown. The young are able to change colour from yellow to green or from green to brown, but only following a moult, not between moults; adults do not change colour. Males have rather protuberant eye tubercles.
Male Palpus: Tibia with ventral, intermediate and retrolateral apophyses. Ventral apophysis rather laterally positioned; stem of apophysis in form of a short, thick finger directed obliquely laterally; apex sharply inclined mesally. Intermediate apophysis obliquely and ventrally directed, branching off at two-thirds the height of retrolateral apophysis and ending obtusely. Retrolateral apophysis terminating in two sharply pointed projections (Figs. 63, 64). Tutaculum rather large, triangular, with pointed tip (Fig. 64). Tegulum quite flat; end of embolus thick, basally downwards directed (Fig. 63).

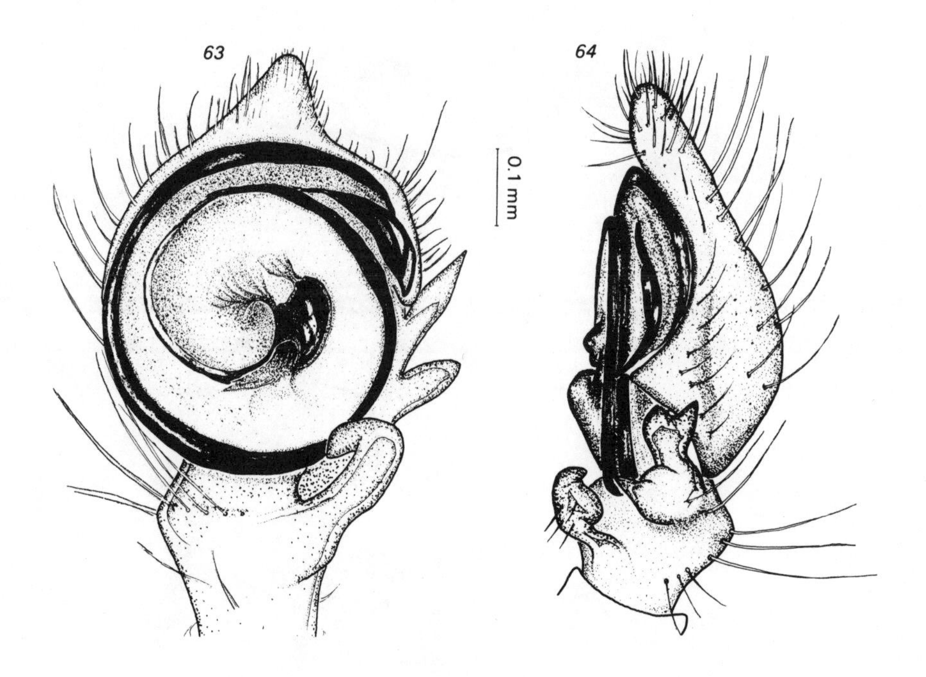

Figs. 63–64: *Heriaeus setiger* (O. P.-Cambridge, 1872); male, left palpus
63. ventral view; 64. lateral view

Female Epigynum: High, convex, hood-shaped structure with transparent walls, opening towards epigastric furrow, on central part of epigynal plate. Distal margins of hood's walls folded or ridged; area between hood and epigastric furrow traversed by series of conspicuous folds. Attachment points of internal spermathecal apodemes are marked by a pair of dark sclerotic circles located near epigastric furrow (Fig. 65; dorsal-inner view, Fig. 66).

Distribution : Southern Europe, northern Africa, Lebanon, Israel.
Israel: Golan Heights (18) to the Carmel Ridge (3), the Coastal Plain (8), Samaria (6) and the Judean Hills (11) to Lahav (10).
Adults in the field are found mainly from May to July.

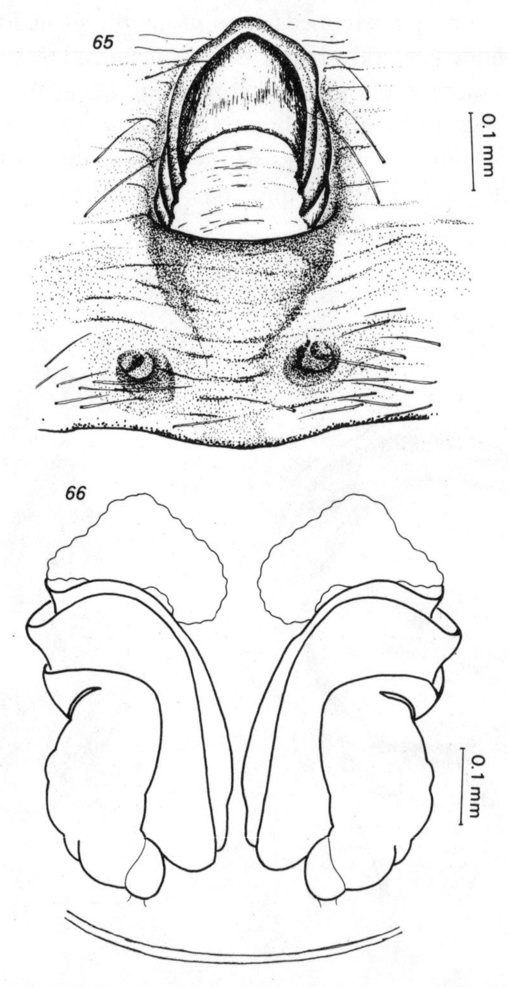

Figs. 65–66: *Heriaeus setiger* (O. P.-Cambridge, 1872); female
65. epigynum; 66. spermathecae, inner view

Heriaeus buffoni (Audouin, 1827)
Figs. 67–70

Thomisus buffonii Audouin, 1827, Explication sommaire des planches d'Arachnides, in: J.C. Savigny, *Description de l'Égypte*, Paris, 22: 396, pl. 6, fig. 10.
Heriaeus buffoni —. Simon, 1875, *Les Arachnides de France*, 2: 206; Roewer, 1954, *Katalog der Araneae*, 2 (1): 864; Bonnet, 1957, *Bibliographia Araneorum*, 2 (3): 2161; Levy, 1973, *Israel J. Zool.*, 22: 138.

Length of male 3.8–4.6 mm, female 5.9 mm. Coloration light yellow to white. Males with low eye tubercles and a very dense cover of long bristles.
Male Palpus: Tibia with distinct ventral and retrolateral apophyses and only a very reduced intermediate apophysis. Ventral apophysis located slightly laterally, with stem in form of a long, bent finger. Retrolateral apophysis rather large, directed slightly ventrally and upwards, slightly pointed at apex (Figs. 67, 68). Tutaculum rounded (Fig. 68). Tegulum rather swollen; tip of embolus stout and recurved (Fig. 67).

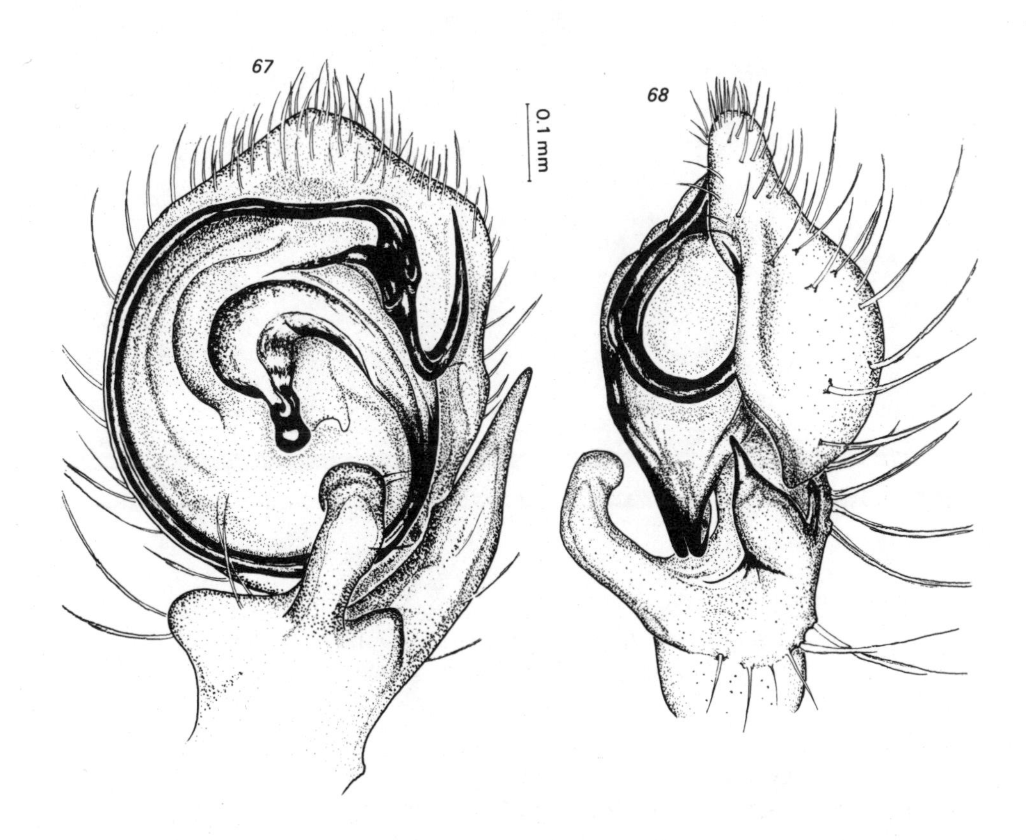

Figs. 67–68: *Heriaeus buffoni* (Audouin, 1827); male, left palpus
67. ventral view; 68. lateral view

Female Epigynum: Sclerotized, rather flat, hood-shaped structure with membranous folds on both sides, located on basal part of epigynal plate; shallow groove extends from beneath hood towards epigastric furrow; attachment points of internal spermathecal apodemes are marked by two dark, conspicuous spots located on distal part of epigynal plate (Fig. 69; dorsal-inner view, Fig. 70).

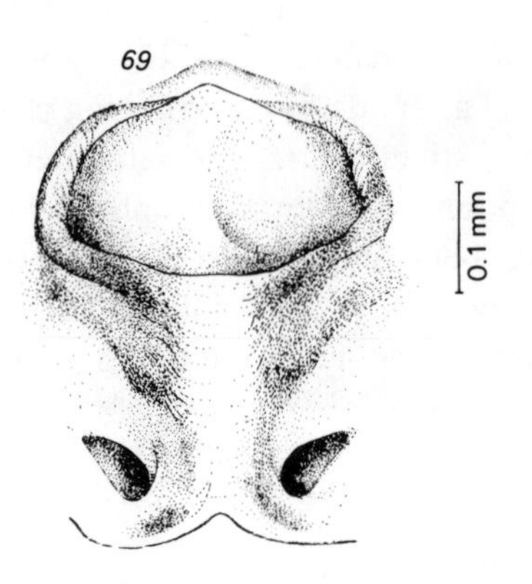

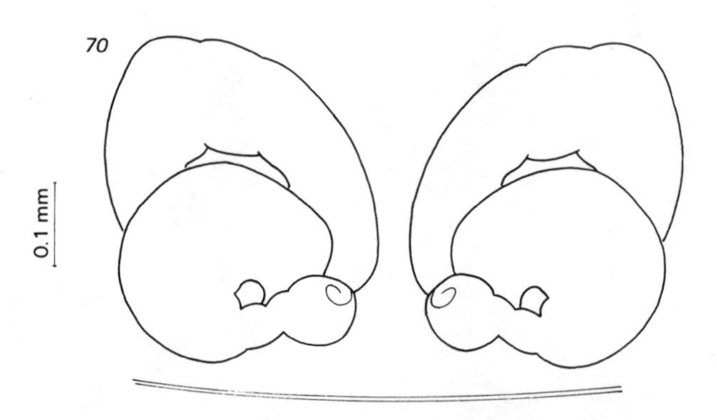

Figs. 69–70: *Heriaeus buffoni* (Audouin, 1827); female
69. epigynum; 70. spermathecae, inner view

Distribution : Egypt, eastern and southern Israel.
Israel : Judean Desert (12), Northern and Central Negev (15, 17).
Adults occur mainly from April to June.

Genus SYNAEMA Simon, 1864

Histoire Naturelle des Araignées, Paris, p. 433

Figs. 71–73

Synema Simon, 1864. Thorell, 1869, *Nova Acta R. Soc. Scient. upsal.*, (3) 7: 36 (emendation).

Type Species: *Aranea globosa* Fabricius, 1775.

Short, rounded spiders (Fig. 71). Carapace truncated anteriorly, very high and convex, as long as wide or slightly longer than wide. Lateral eyes borne on separate tubercles; tubercles of approximately the same size. Anterior row of eyes slightly recurved or straight and equidistant (Figs. 72, 73). Posterior row of eyes rather recurved, equidistant, or distinctly farther from each other than from posterior-lateral eyes. Lateral eyes larger than median eyes. Median ocular quadrangle wider than long, narrow anteriorly and wide posteriorly (Fig. 72). Legs short; two anterior pairs longer than two posterior pairs; claws with distinct denticles. Opisthosoma oval or rounded (Fig. 71).

Synaema species have brilliant colours. The dorsal surface of the opisthosoma usually bears a white or black pattern resembling the Cedar of Lebanon, conspicuously set on a black, white or occasionally red or yellow background. Light patches often occur on venter. These spiders live in vegetation, often on the lurk inside flowers. They move with agility, catching their prey with a few rapid strides.

About 130 species distributed throughout the world. Three species in Israel.

Key to the Species of Synaema in Israel

Males:

1. Tibia of palpus distinctly longer than wide (Fig. 74). — **S. globosum** (Fabricius)
– Tibia of palpus short, about as long as wide (Figs. 78, 82) — 2
2. Retrolateral tibial apophysis of palpus with truncated tip bearing a pointed, triangular protuberance (Figs. 78, 79). — **S. plorator** (O. P.-Cambridge)
– Retrolateral tibial apophysis of palpus with step-like recess on inner side at middle, following part pointed and slightly inclined (Figs. 82, 83). — **S. diana** (Audouin)

Females:

1. Prosoma green. Epigynum traversed by a very elongated sclerotic structure with side-corners connected to each other by a transparent membrane (Fig. 84). — **S. diana** (Audouin)
– Prosoma brown to dark brown. Epigynum traversed by a sclerotic structure without a transparent membrane (Figs. 76, 80) — 2
2. Structure on epigynum slender and elongate: more than four times wider than its height at narrow middle (Fig. 76). Spermatheca formed of small, tight coils throughout (Fig. 77, require dissection). — **S. globosum** (Fabricius)

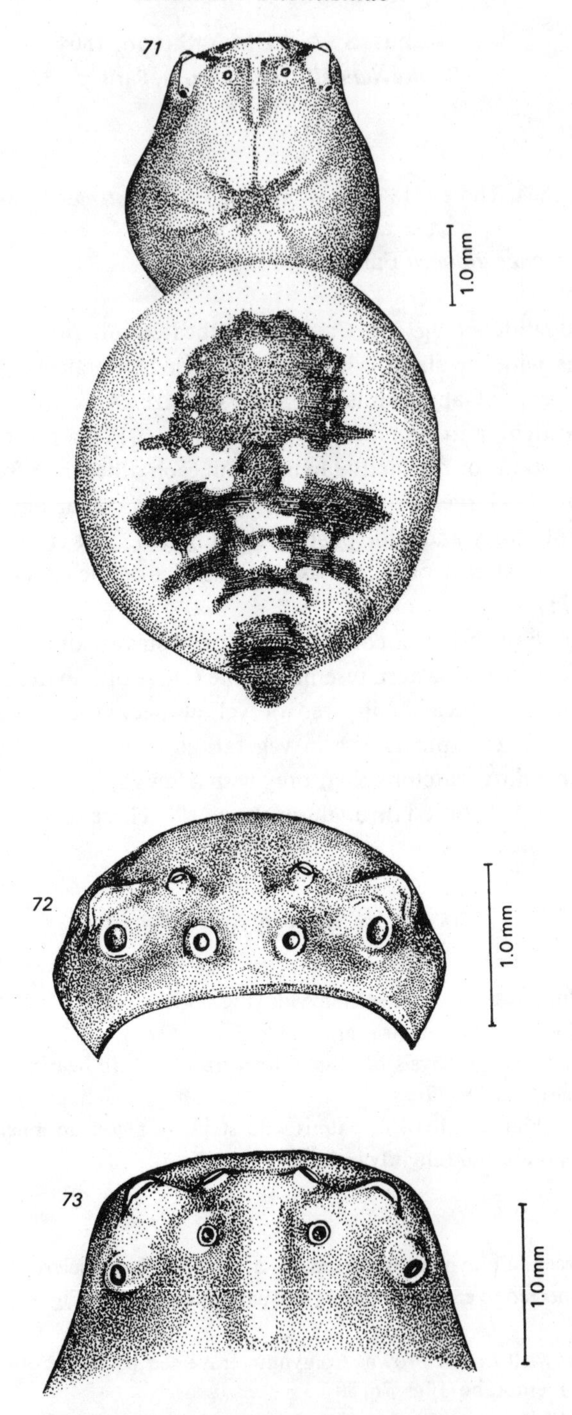

Figs. 71–73: *Synaema* Simon, 1864; female
71. dorsal view of spider; 72. carapace, frontal view;
73. carapace, dorsal view of anterior part

Structure on epigynum broad: less than three times wider than its height at narrow middle (Fig. 80). Spermatheca formed of small coils becoming larger gradually (Fig. 81, require dissection). **S. plorator** (O. P.-Cambridge)

Synaema globosum (Fabricius, 1775)
Figs. 74–77

Aranea globosa Fabricius, 1775, *Systema Entomologiae*, Flensburg & Lipsiae, p. 432.
Synaema globosum —. Simon, 1875, *Annls Soc. ent. Fr.*, (5) 5 *Bull.*: 197; Roewer, 1954, *Katalog der Araneae*, 2 (1): 885; Bonnet, 1958, *Bibliographia Araneorum*, 2 (4): 4207; Levy, 1975, *Israel J. Zool.*, 24: 157.

Length of male 4.0–5.7 mm, female 5.8–7.3 mm. Coloration of prosoma dark brown. Opisthosoma of female with a dark pattern on dorsal surface, usually traversed by light bands; venter usually with a large white patch below epigynum, sometimes with additional white spots. Specimens with red and yellow colourings are occasionally found.
Male Palpus: Relatively small. Tibia distinctly longer than wide (Figs. 74, 75). Tibial apophyses parallel, directed laterally and obliquely. Ventral apophysis finger-like, evenly thick and with a rounded tip. Retrolateral apophysis with step-like recess at middle on inner side; following part needle-like, pointed (Fig. 74). Embolar duct encircles bulb about one and a half times, ending apically in slightly bent tip (Fig. 74).

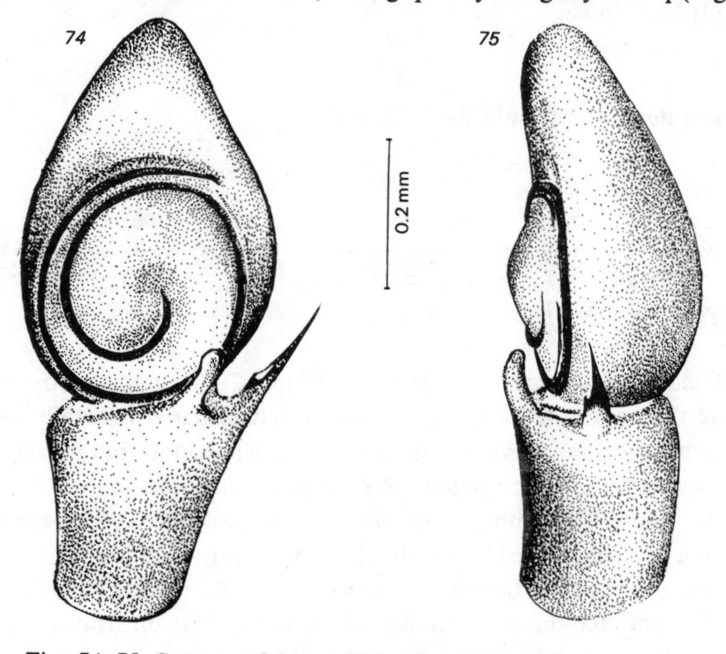

Figs. 74–75: *Synaema globosum* (Fabricius, 1775); male, left palpus
74. ventral view; 75. lateral view

Female Epigynum: Relatively small. Epigynal plate oval, centrally traversed by an elongated, slender, sclerotic structure (Fig. 76). Width of central structure more than four times the height at narrow middle (Fig. 76; dorsal-inner view, Fig. 77).

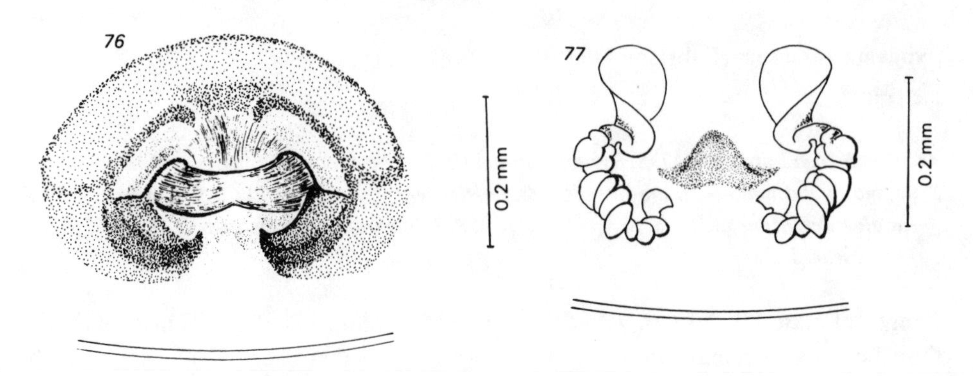

Figs. 76–77: *Synaema globosum* (Fabricius, 1775); female
76. epigynum; 77. spermathecae, inner view

Distribution: Throughout the Palaearctic region.
Israel: From the Golan Heights (18) throughout the mountains to the Judean Hills (11) and the Coastal Plain to Be'er Toviyya (9).
Adult males are found from April to June, mainly in May, and females from April to July, mainly May–June.

Synaema plorator (O. P.-Cambridge, 1872)
Figs. 78–81

Thomisus plorator O. P.-Cambridge, 1872, *Proc. zool. Soc. Lond.*, p. 306.
Synaema plorator —. Simon, 1884, *Annls Soc. ent. Fr.*, (6) 4: 322; Roewer, 1954, *Katalog der Araneae*, 2 (1): 886; Bonnet, 1958, *Bibliographia Araneorum*, 2 (4): 4214; Levy, 1975, *Israel J. Zool.*, 24: 159.

Length of male 4.3–4.5 mm, female 6.2–9.0 mm. Coloration of prosoma usually brown. Opisthosoma of female with a continuous dark pattern dorsally, not usually interrupted by white bands; venter usually without white spots, but occasionally with spots. Specimens with red and yellow colourings are sometimes found.
Male Palpus: Large. Tibia short, more or less as long as wide. Ventral tibial apophysis large, thick, oblique, bent laterally (Fig. 78); stem at middle curved outwards (Fig. 79); end of apophysis bending mesally (Fig. 78). Retrolateral apophysis forms a thick lateral continuation of distal side of tibia (Fig. 78); tip truncated, bearing a pointed, triangular protuberance (Fig. 79). Embolar duct encircles bulb more than twice, ending apically in hook-like, recurved tip (Fig. 78).

Female Epigynum: Large. Epigynal plate rounded, with broad, transverse, central sclerotic structure (Fig. 80). Width of central structure less than three times the height at narrow middle (Fig. 80; dorsal-inner view, Fig. 81).

Distribution: Mediterranean countries, southern Europe to Central Asia.
Israel: Central Coastal Plain (8), Foothills of Judea (10) and the Judean Hills (11).
Males are found in March, adult females from February to May.

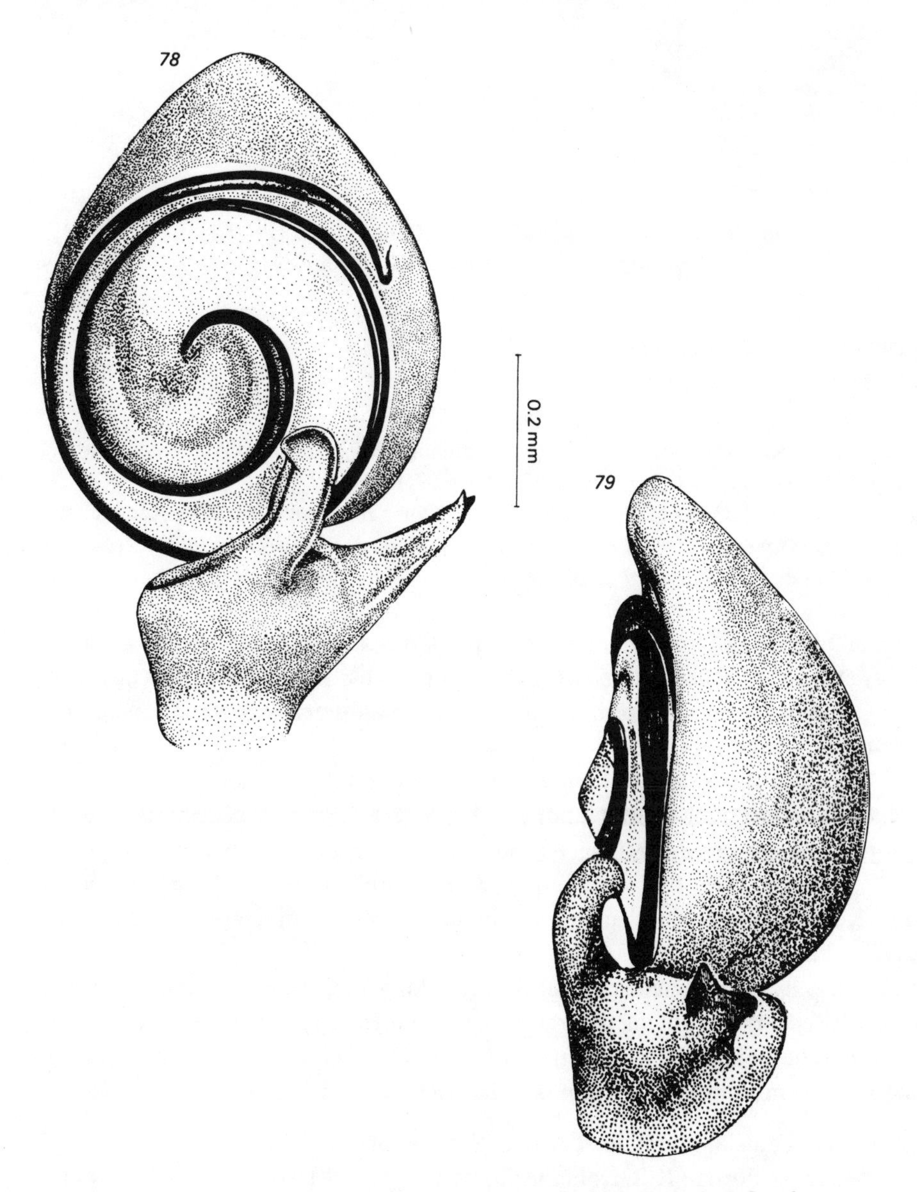

Figs. 78–79: ***Synaema plorator*** **(O. P.-Cambridge, 1872); male, left palpus**
78. ventral view; 79. lateral view

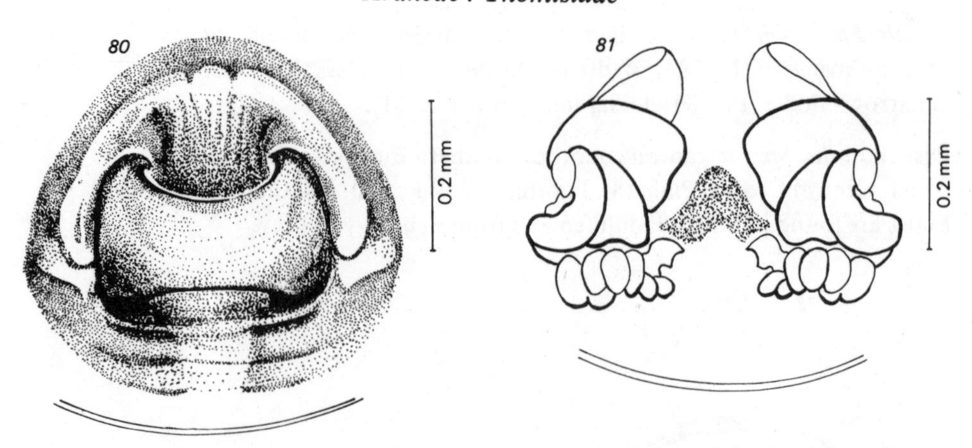

Figs. 80–81 : *Synaema plorator* (O. P.-Cambridge, 1872); female
80. epigynum; 81. spermathecae, inner view

Synaema diana (Audouin, 1827)
Figs. 82–85

Thomisus diana Audouin, 1827, Explication sommaire des planches d'Arachnides, in: J. C. Savigny, *Description de l'Égypte*, Paris, 22 : 399, pl. 7, figs. 1, 2.

Synaema diana —. Simon, 1882, *Annali Mus. civ. Stor. nat. Genova*, 18: 224; Roewer, 1954, *Katalog der Araneae*, 2 (1): 886 (as *S. audouini*); Bonnet, 1958, *Bibliographia Araneorum*, 2 (4): 4205; Levy, 1975, *Israel J. Zool.*, 24: 161.

Length of male 3.6–4.7 mm, female 4.4–6.5 mm. Coloration of prosoma deep green with white eye tubercles. Opisthosoma of female white with black and red spots and a light brown Cedar of Lebanon pattern; opisthosoma of male green with white patches. Legs green with red rings.

Male Palpus: Of medium size. Tibia short, approximately as long as wide. Tibial apophyses oblique, bent laterally, not parallel (Fig. 82). Ventral apophysis large with tip projecting slightly mesally (Fig. 82). Retrolateral apophysis with step-like recess at middle on inner side, the following section being pointed and slightly inclined (Figs. 82, 83). Embolar duct encircles bulb about twice, ending apically in a distinct bend (Fig. 82).

Female Epigynum: Epigynal plate with a very elongated, transverse central sclerotic structure (Fig. 84). Width of central structure more than five times the height at narrow middle (Fig. 84); side-corners of structure connected to each other by a transparent membrane (Fig. 84; dorsal-inner view, Fig. 85).

Distribution: Tunisia, Egypt, East Africa, Yemen, Israel.

Israel: Along the Jordan Rift, from the Golan Heights (18) to the 'Arava Valley (14).

Synaema diana spins a small retreat web on tips of *Tamarix* twigs. Adults are found almost throughout the year, except possibly from October to December.

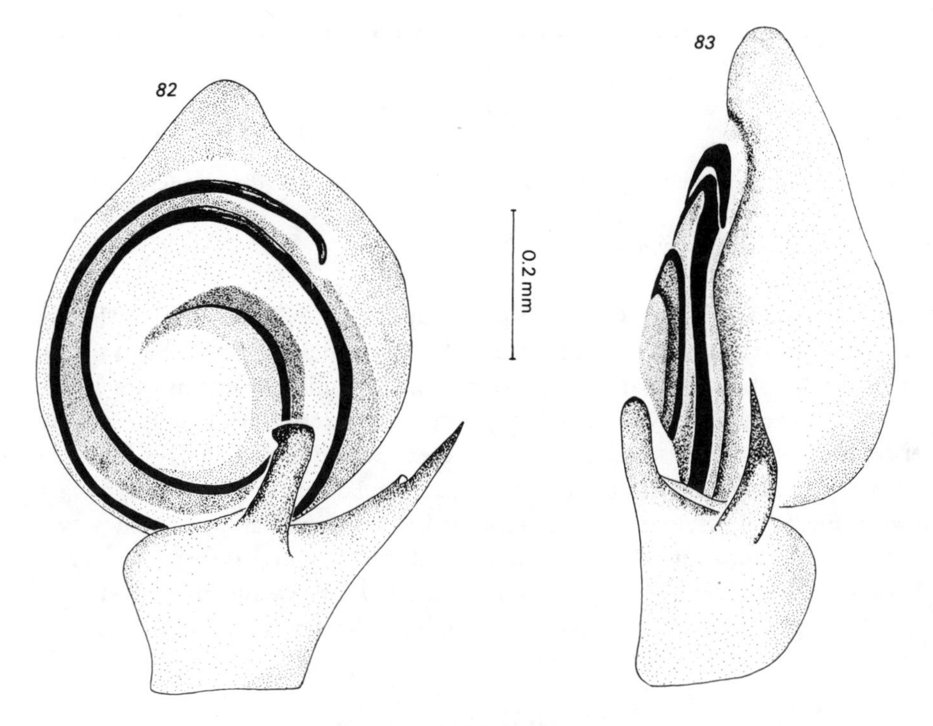

Figs. 82–83: *Synaema diana* (Audouin, 1827); male, left palpus
82. ventral view; 83. lateral view

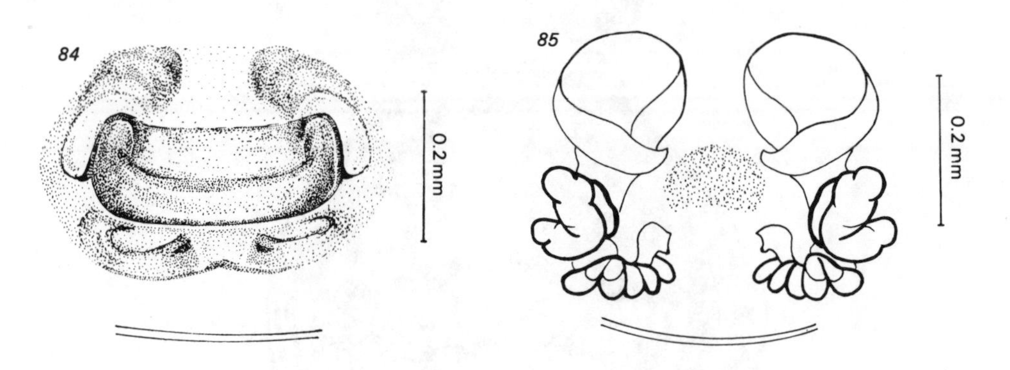

Figs. 84–85: *Synaema diana* (Audouin, 1827); female
84. epigynum; 85. spermathecae, inner view

Genus OXYPTILA Simon, 1864

Histoire Naturelle des Araignées, Paris, pp. 439, 527

Figs. 86–88

Ozyptila Simon, 1864. Thorell, 1869, *Nova Acta R. Soc. Scient. upsal.*, (3) 7: 36 (emendation).

Type Species: *Thomisus brevipes* Hahn, 1826.

Short, stout spiders (Fig. 86). Carapace pear-shaped (Fig. 87), rather high, slightly convex and approximately as long as wide; clypeus vertical (Fig. 88). Lateral eyes borne on low, distinctly separated tubercles (Fig. 87); anterior-lateral eyes larger than all other eyes; both eye-rows slightly recurved; anterior-median eyes usually closer to each other than to anterior-lateral eyes; posterior-median eyes closer to each other than to posterior-lateral eyes; median ocular quadrangle usually longer than wide (Fig. 88). Legs, mainly two anterior pairs, short, thick and with spines; two anterior pairs of approximately same length and distinctly longer than two posterior pairs; claws with few denticles. Opisthosoma narrow anteriorly and broad posteriorly (Fig. 86), dorsally flat and covered by thick and thin bristles; bristles partly clavate or spatulate.

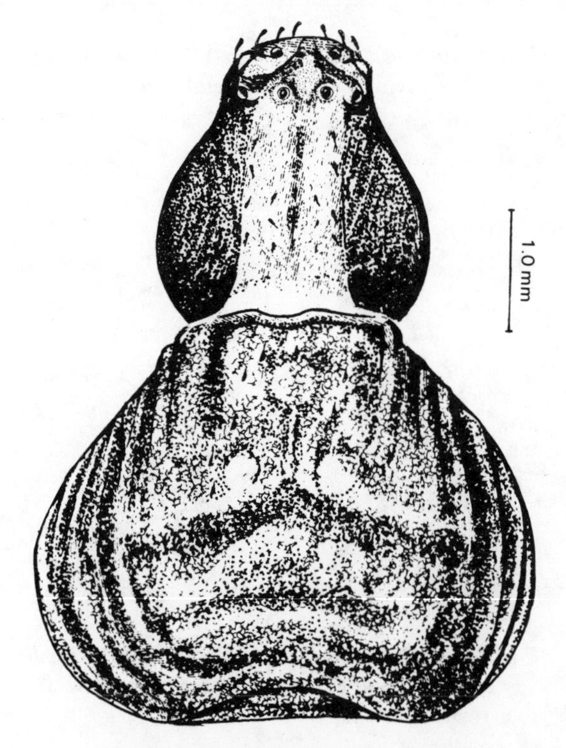

Fig. 86: *Oxyptila* Simon, 1864; female, dorsal view of spider

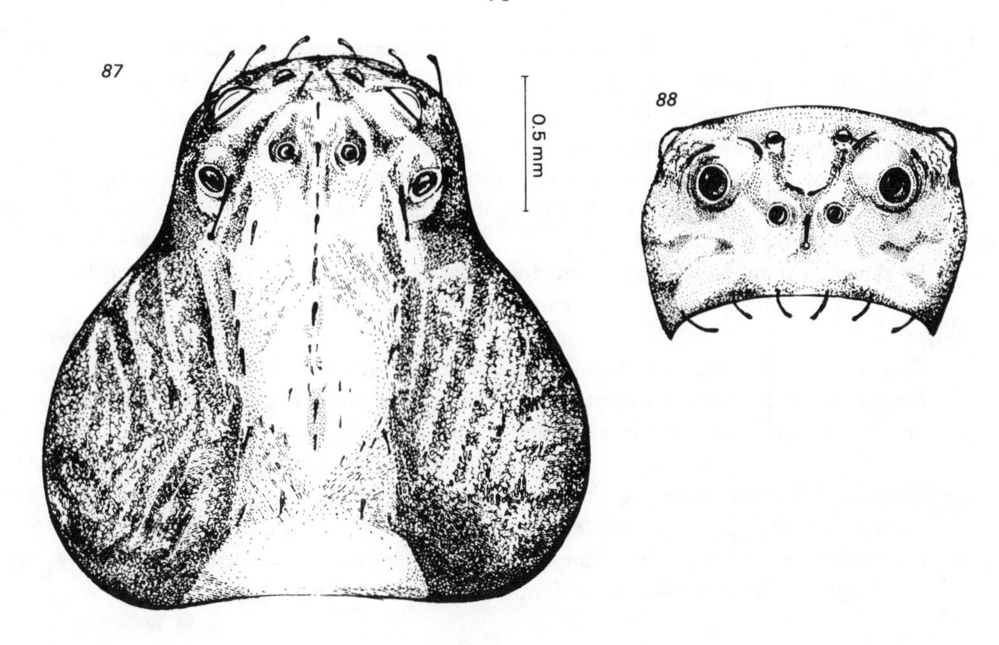

Figs. 87–88: *Oxyptila* Simon, 1864; female
87. carapace, dorsal view; 88. carapace, frontal view

Oxyptila spiders live under stones or the bark of trees and sometimes on plants and low bushes. They move slowly and are rather inconspicuous owing to their dark brown coloration, which often matches that of the background. Prey is caught by ambush.

More than 100 species distributed throughout the world. Many are known from one sex only, as the males and females cannot be matched with certainty unless a pair is captured together or the spiders are bred from eggs to maturity. Seven species in Israel, four represented only by females and two only by males.

Key to the Species of Oxyptila in Israel

Males:

1. Tibia of palpus with three apophyses, the ventral one hammer-like with thick crossbars (Figs. 98, 99). Tegulum of palpus bears a fine, looped stylus (Fig. 98). **O. complicata** Levy
- Tibia of palpus with only two apophyses, the ventral one not hammer-headed. Tegulum of palpus bears other structures 2
2. Retrolateral tibial apophysis of palpus bears a very large, pointed dark spine on the outside (Figs. 89, 90). Tegulum of palpus with three crescentic folds. Tip of embolus is split into two fine stylets (Figs. 89, 90). **O. tricoloripes** Strand
- Retrolateral tibial apophysis of palpus without an outer spine. Tegulum of palpus with two concave protuberances (Figs. 102, 103). Tip of embolus in form of a fine, arched stylet (Fig. 102). **O. aradensis** Levy

Females :

1. Epigynum with a distinct central protrusion rising close to epigastric furrow and pointing towards prosoma (Figs. 91, 104) 2
- Epigynum without such a protrusion 3
2. Epigynum with a large, sclerotic, hood-shaped structure which opens towards large central protrusion (Fig. 91). Dark, thick coils of spermathecae usually visible on both sides of central protrusion (Fig. 91). **O. tricoloripes** Strand
- Epigynum without hood-shaped structure. Flanks of central protrusion covered by a transverse fold and spermathecae not visible through integument (Fig. 104). **O. omega** Levy
3. Epigynum with distinct, basal, hood-shaped structure (Figs. 96, 100) 4
- Epigynum wihout hood-shaped structure. Two dark arches join at middle of epigynum forming a raised median septum extending into a sclerotized funnel (Fig. 94). **O. clavigera** (O. P.-Cambridge)
4. Hood-shaped structure on epigynum small, rounded, with a circular opening (Fig. 96). Central part of epigynum surrounded by two large folds. **O. rigida** (O. P.-Cambridge)
- Hood-shaped structure on epigynum large, tent-shaped with a wide, triangular opening; distal edges of hood's opening coil to form dark, thick, tight loops (Fig. 100). **O. judaea** Levy

Oxyptila tricoloripes Strand, 1913

Figs. 89–93

Thomisus confluens (C. L. Koch, 1845). O. P.-Cambridge, 1872, *Proc. zool. Soc. Lond.*, p. 305 (misdetermination).

Oxyptila rigida (O. P.-Cambridge, 1872). Kulczyński, 1908, *Bull. Acad. Sci. Cracovie*, p. 70, pl. 2, fig. 15; Roewer, 1954, *Katalog der Araneae*, 2(1): 879; Bonnet, 1958, *Bibliographia Araneorum*, 2 (4): 3262 (misdetermination).

Oxyptila tricoloripes Strand, 1913, *Arch. Naturgesch.*, 79: 152; Roewer, 1954, *Katalog der Araneae*, 2 (1): 880; Bonnet, 1958, *Bibliographia Araneorum*, 2 (4): 3266; Levy, 1975, *Israel J. Zool.*, 24: 167; Fet, 1982, *Trans. Acad. Sci. Turkmen. SSR*, 5: 74.

Oxyptila pickardi Levy, 1975, *Israel J. Zool.*, 24: 165.

Length of male 3.2–3.3 mm, female 4.3–6.8 mm. Coloration of prosoma dark brown in male, light-coloured on dorsum in female. Opisthosoma dark brown with a fine white line on frontal margin in male, and with light scattered spots in female. Anterior pairs of legs of male dark brown, only the tips light-coloured; posterior pairs very light brown; legs of female light brown mottled with dark patches.

Male Palpus: Relatively small. Tibia with ventral and retrolateral apophyses. Ventral apophysis with broad stem, concave on inside (Fig. 90); apex bent slightly inwards and mesally (Fig. 89). Retrolateral apophysis large, concave at base and ending in pointed, black tip; apophysis on round lateral socket, bearing a large, cone-shaped, pointed, black spine (Figs. 89, 90). Tegulum wih three crescentic folds, progressively increasing in size, with flanks slightly overlapping one another (Fig. 89). Embolar end split into two fine stylets, one pointing towards tegulum, the other extending over edges of tutaculum and bending downwards (Figs. 89, 90).

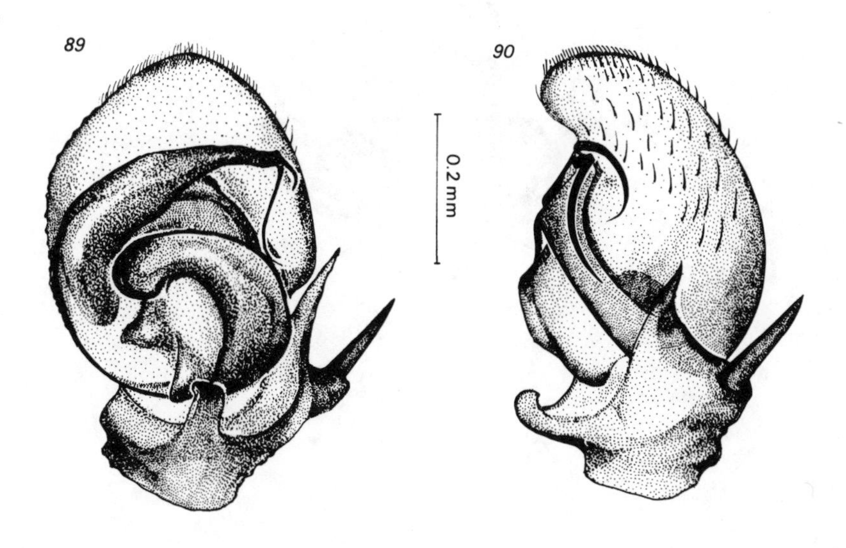

Figs. 89–90: *Oxyptila tricoloripes* Strand, 1913; male, left palpus
89. ventral view; 90. lateral view

Female Epigynum: Basal part of epigynal plate wih large, sclerotic hood-shaped structure (Fig. 91); hood with narrow, slit-like opening directed towards epigastric furrow. Distal part of epigynal plate with large, rectangular protrusion, basally directed, extending close to hood's opening; apex of protrusion intact (Fig. 92) or occasionally with a deep, longitudinal split (Fig. 91; dorsal-inner view, Fig. 93).

Distribution: Near Caspian Sea, USSR (Kopetdagh, Baku), Cyprus, Israel.
Israel: From the Golan Heights (18), Lower Galilee (2) and Lake Kinneret (7) to the Judean Hills (11), Foothills of Judea (10) and the Central Coastal Plain (8).

Adults of both sexes were collected together at mountain woods in southwestern Turkmenia (Fet, 1982) proving thereby the matching formerly predicted of *O. pickardi* females with males of *O. tricoloripes* (Levy, 1975: 168). In Israel, adult males are found only in November, while adult females are found throughout winter and spring, October to April. In Turkmenia, *O. tricoloripes* is likewise considered an autumn species: males occur only from September to November, and females are found until April–May. In Israel, females crouching on a disc-shaped egg sac were collected in March and April; the young develop and moult throughout summer.

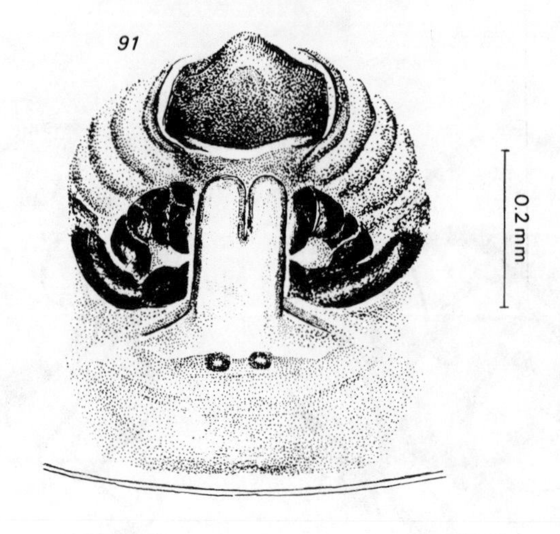

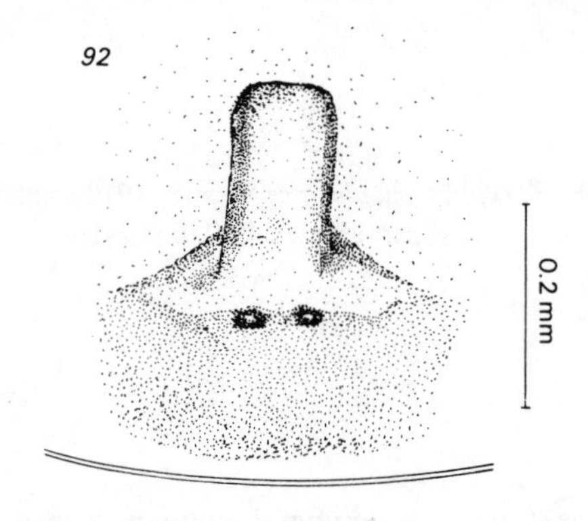

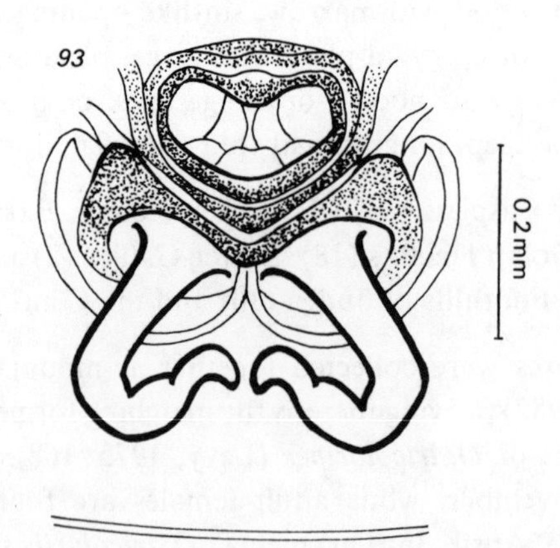

Figs. 91–93: *Oxyptila tricoloripes* Strand, 1913; female
91. epigynum; 92. intact rectangular protrusion of epigynum (detail);
93. spermathecae, inner view

Oxyptila clavigera (O. P.-Cambridge, 1872)
Figs. 94, 95

Thomisus claviger O. P.-Cambridge, 1872, *Proc. zool. Soc. Lond.*, p. 306.
Oxyptila clavigera —. Simon, 1875, *Les Arachnides de France*, 2: 241; Roewer, 1954, *Katalog der Araneae*, 2 (1): 876; Bonnet, 1958, *Bibliographia Araneorum*, 2 (4): 3254.

Male unknown. Length of female 3.8–4.2 mm. Coloration of preserved specimens: prosoma dark brown on sides and light-coloured on dorsum. Opisthosoma yellowish, mottled with slightly darker patches. Legs brown, mottled with light spots.
Female Epigynum: Two dark, sclerotic arches on basal part of epigynal plate join at middle and form a raised median septum. Median septum extends only about half way to epigastric furrow, stopping in front of fine, transverse, funnel-shaped concavity with dark sclerotic rims (Fig. 94; dorsal-inner view, Fig. 95).

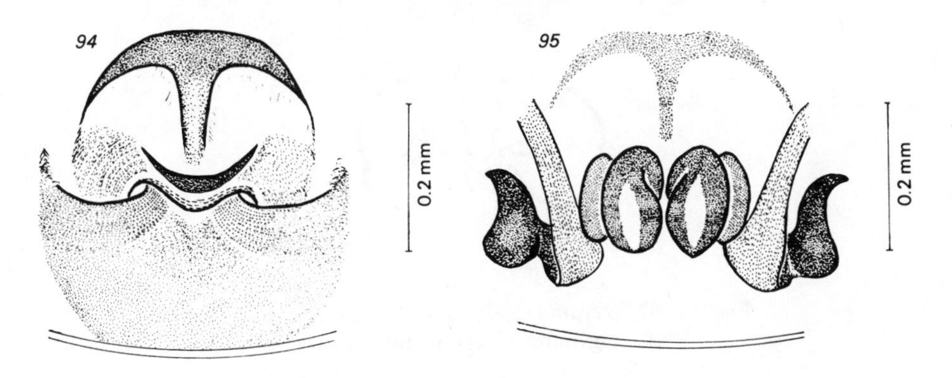

Figs. 94–95: *Oxyptila clavigera* (O. P.-Cambridge, 1872); female
94. epigynum; 95. spermathecae, inner view

Distribution, Israel: Known so far only from two type specimens found by O. P.-Cambridge (1872: 306), one on the plains of the Jordan (13) and another at Nazareth (2), under stones (? April); (deposited in the Hope Entomological Collections, University Museum, Oxford, England; B. 1256, t. 25).

Oxyptila rigida (O. P.-Cambridge, 1872)
Figs. 96, 97

Thomisus rigidus O. P.-Cambridge, 1872, *Proc. zool. Soc. Lond.*, p. 305.
Oxyptila rigida —. Levy, 1975, *Israel J. Zool.*, 24: 168 (not *O. rigida* —. Kulczyński, 1908).

Male unknown. Length of female 4.8 mm. Coloration of opisthosoma dark brown mottled with light patches. Prosoma darker on sides than dorsally.
Female Epigynum: Basal part of epigynal plate with small, rounded, hood-shaped

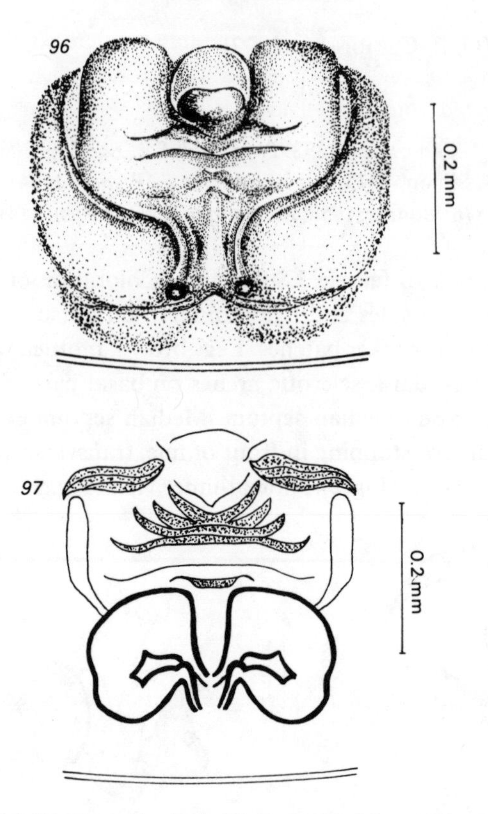

Figs. 96–97: *Oxyptila rigida* (O. P.-Cambridge, 1872); female
96. epigynum; 97. spermathecae, inner view

structure set in a deep depression (Fig. 96); hood with light-coloured, transparent walls and an almost circular opening directed towards epigastric furrow. Central part of epigynum traversed by a few ridges and surrounded on both sides by two large folds (Fig. 96; dorsal-inner view, Fig. 97).

Distribution, Israel: Near Lake Kinneret (7) and Dead Sea area (13), along the Jordan Rift.
Only two specimens have been collected so far. The female type was apparently found in April, the other, near the Kinneret, in May.

Oxyptila complicata Levy, 1975
Figs. 98, 99

Oxyptila complicata Levy, 1975, *Israel J. Zool.*, 24: 169.

Length of male 3.0–3.2 mm; female unknown. Coloration of prosoma dark brown, opisthosoma light-coloured with scattered black spots.

Male Palpus: Tibia with ventral, intermediate and retrolateral apophyses. Ventral apophysis hammer-like with an upper, dark, hood-shaped extension and an oblique, downwards directed thick extension (Figs. 98, 99). Intermediate apophysis stout, slightly concave inside and covered with dot-like depressions on outside (Fig. 99). Retrolateral apophysis large, closely attached to bulb and separated by a wide concavity from the intermediate apophysis (Fig. 99). Tegulum on lower part with large, black, downwards inclined tegular apophysis with pointed tip bending sharply mesally (Fig. 98); tegulum on upper part bearing fine, sclerotic, whip-like stylus (not the embolus!) bending downwards and forming a loop with end pointing upwards (Fig. 98).

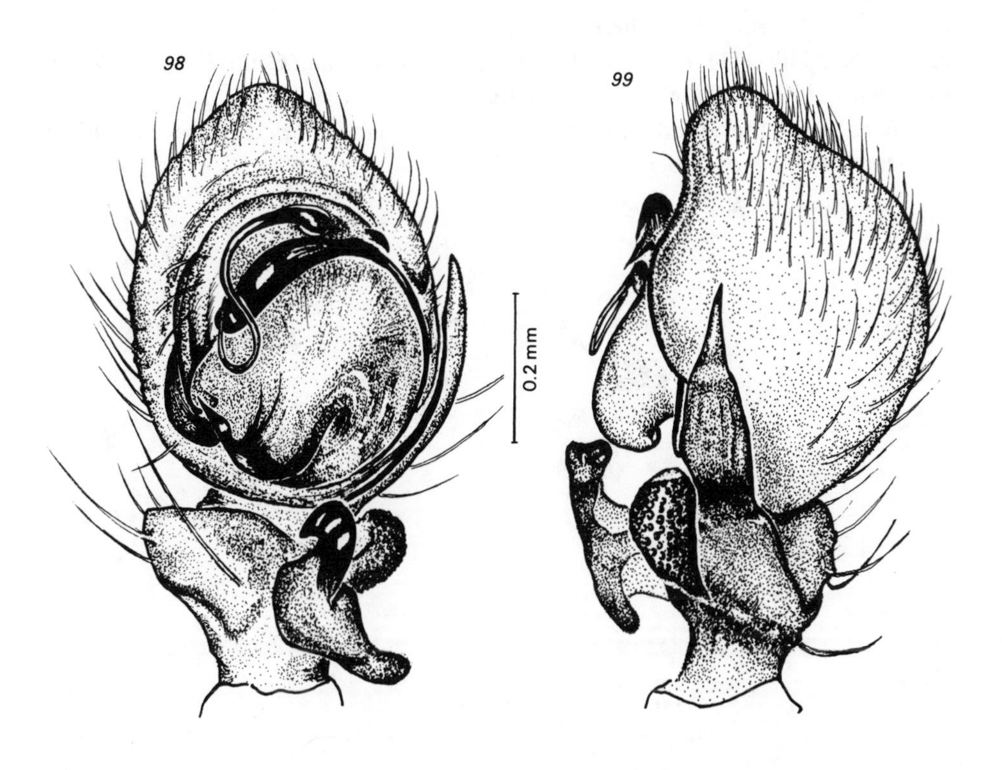

Figs. 98–99: *Oxyptila complicata* Levy, 1975; male, left palpus
98. ventral view; 99. lateral view

Distribution, Israel: Deganya, near Lake Kinneret (7) and Dead Sea area (13), along the Jordan Rift.
Adult males were found in January and May. *Oxyptila complicata* possibly represents the matching male of *O. rigida*, but direct evidence is still lacking.

Oxyptila judaea Levy, 1975
Figs. 100, 101

Oxyptila judaea Levy, 1975, *Israel J. Zool.*, 24: 171.

Male unknown. Length of female 5.7–9.5 mm. Coloration light brown mottled with dark brown spots.
Female Epigynum: Basal part of epigynal plate with large, tent-like structure having a wide, triangular opening directed downwards (Fig. 100); a dark, thick, sclerotized band extends along the tent's edges and opening, forming thick, convoluted loops at lowest corners of tent's opening (Fig. 100; dorsal-inner view, Fig. 101).

Distribution: Central and southern Israel, Sinai.
Israel: Judean Hills (11), Foothills of Judea (10), Northern and Central Negev (15, 17).
Sinai: South of Elat and in the Sinai Mountains (22).
Adult females are found throughout the year; in the laboratory, eggs have been laid from September to November.

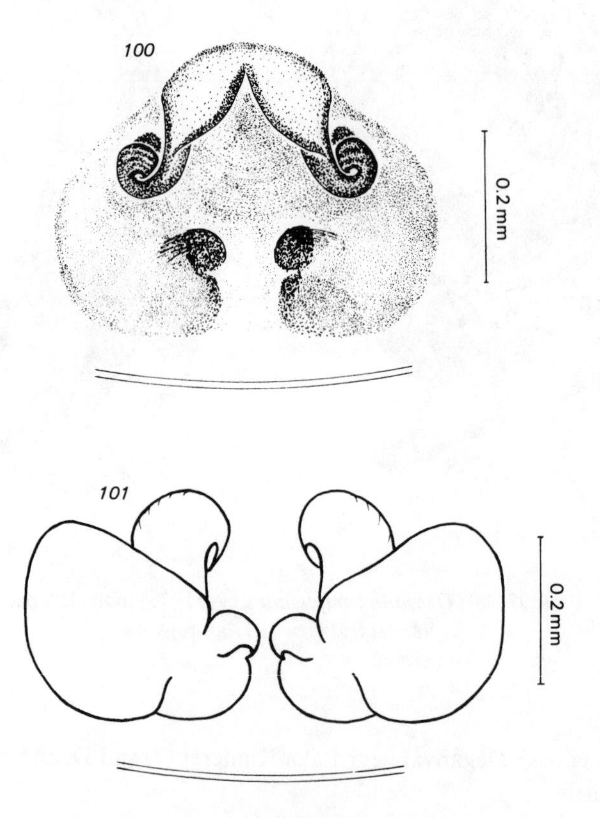

Figs. 100–101: *Oxyptila judaea* Levy, 1975; female
100. epigynum; 101. spermathecae, inner view

Oxyptila aradensis Levy, 1975
Figs. 102, 103

Oxyptila aradensis Levy, 1975, *Israel J. Zool.*, 24: 172.

Length of male 4.1–4.7 mm; female unknown. Coloration of opisthosoma whitish-grey with dots and dark, transverse bands. Prosoma brown with light-coloured spots. Legs brown mottled with white.
Male Palpus: Tibia with ventral and retrolateral apophyses. Ventral apophysis thumb-like, directed slightly outwards and with round apical swelling bending mesally (Fig. 102). Retrolateral apophysis thick, with a small step-like bulge on inner side, close to tip (Fig. 102). Transparent tutaculum inconspicuous (Figs. 102, 103). Tegulum on medial upper part bearing sclerotic concave protuberance ending in black pointed tip (Figs. 102, 103); upper edges of protuberance slightly cover a cup-like fold with an outwards directed opening (Figs. 102, 103).

Distribution, Israel: Judean Hills (Jerusalem; 11) and Central Negev ('Arad; 17). Adult males occur in June and July. The matching female might be represented by *O. judaea*.

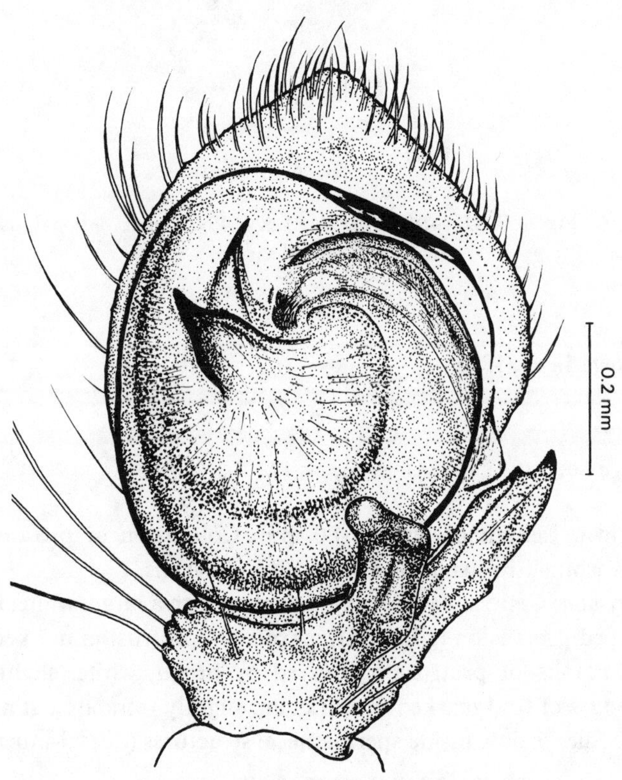

Fig. 102: *Oxyptila aradensis* Levy, 1975; male, left palpus, ventral view

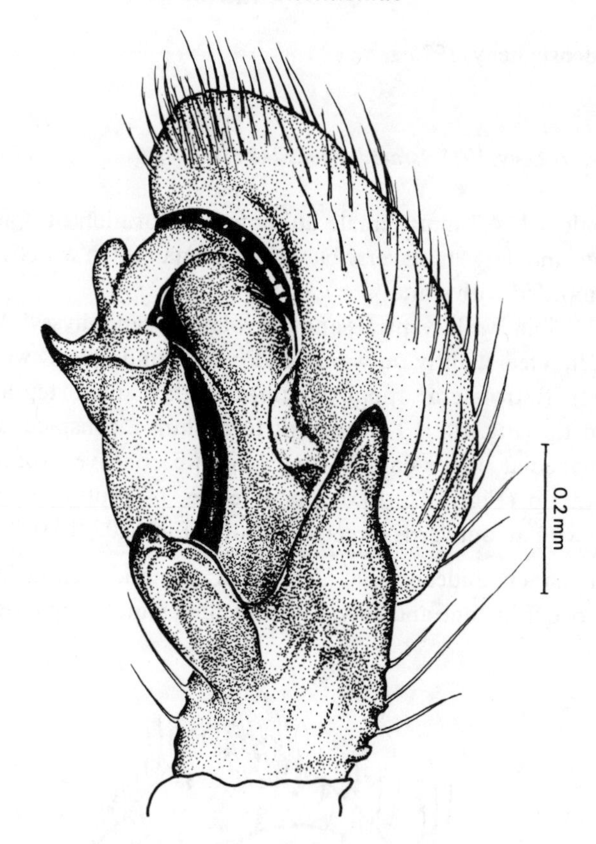

Fig. 103: *Oxyptila aradensis* Levy, 1975; male, left palpus, lateral view

Oxyptila omega Levy, 1975
Figs. 104, 105

Oxyptila omega Levy, 1975, *Israel J. Zool.*, 24: 172.

Male unknown. Length of female 4.0 mm. Coloration of prosoma dark brown; opisthosoma whitish-brown traversed by dark chevrons.
Female Epigynum: Central part of epigynal plate with a large, white, basally pointing, 'omega'-shaped protrusion (Fig. 104); margins of protrusion marked by brown sclerotic band. Flanks of protrusion covered by broad, white, slightly elevated fold (Fig. 104); edges of fold marked by dark rim, slightly retracting at middle (Fig. 104). Small oval bodies visible inside spermathecal structures (dorsal-inner view, Fig. 105).

Distribution, Israel : Valley of Yizre'el (Bet Alfa ; 5).
An adult female was collected in December.

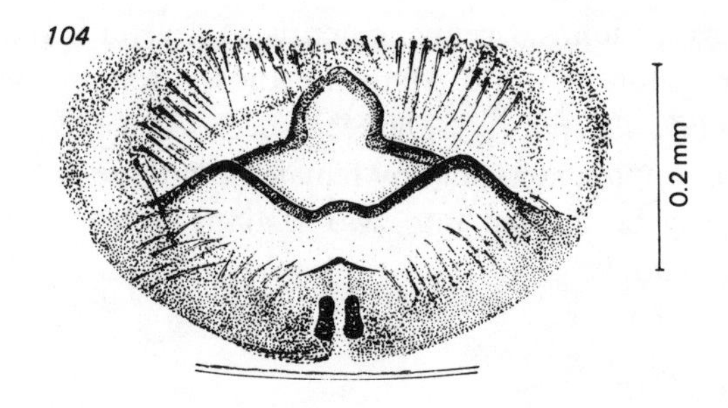

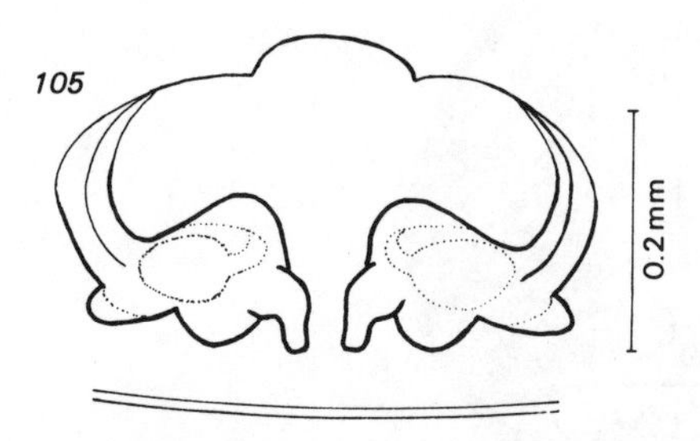

Figs. 104–105: *Oxyptila omega* Levy, 1975; female
104. epigynum; 105. spermathecae, inner view

Genus XYSTICUS C. L. Koch, 1835

Arachniden, in : Panzer, *Faunae Insectorum Germaniae initia*,
Regensburg, Heft 129, pl. 16, 17

Figs. 106–108

Type Species: *Araneus cristatus* Clerck, 1757.

Squat, short spiders (Fig. 106). Carapace truncated anteriorly, moderately high, slightly convex above and approximately as long as wide (Figs. 107, 108). Lateral eyes on separate, low tubercles, distinctly larger than median eyes (Figs. 107, 108); both eye-rows slightly recurved. Anterior-median eyes usually farther from each other than from anterior-lateral eyes (Fig. 107); posterior-median eyes usually closer to each other than to posterior-lateral eyes or eyes equidistant (Fig. 108); median ocular quadrangle nearly square. Legs short, stout and usually spiny; two anterior pairs longer and thicker than two posterior pairs; claws with denticles. Opisthosoma flattened above, widest behind middle (Fig. 106), with short, blunt or fine, pointed bristles.

Xysticus spiders are usually found under stones or bark of trees. They are heavy bodied, slow moving and lie in ambush for their prey. Their coloration is generally dull yellowish-brown, but males occasionally are more conspicuously marked.

Xysticus is the largest thomisid genus, comprising several hundred species, distributed throughout most of the world. Many are known from one sex only, either male or female, as in *Oxyptila* species. Seventeen species in Israel: ten are known from both sexes, four are represented only by females and three only by males. The local *Xysticus* species may be divided into two sections based on the type of bristles covering the spider's body.

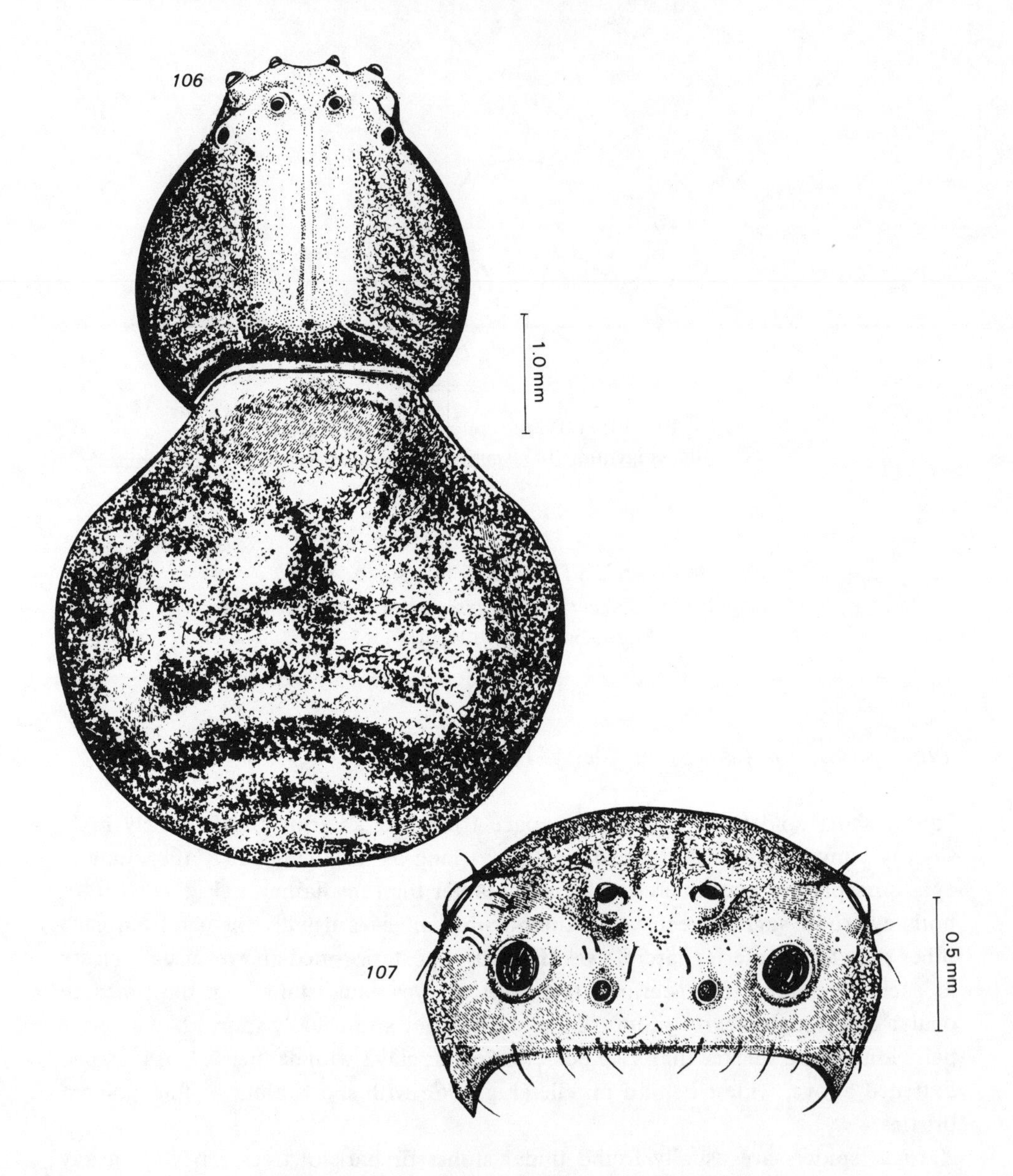

Figs. 106–107: *Xysticus* C. L. Koch, 1835; female
106. dorsal view of spider; 107. carapace, frontal view

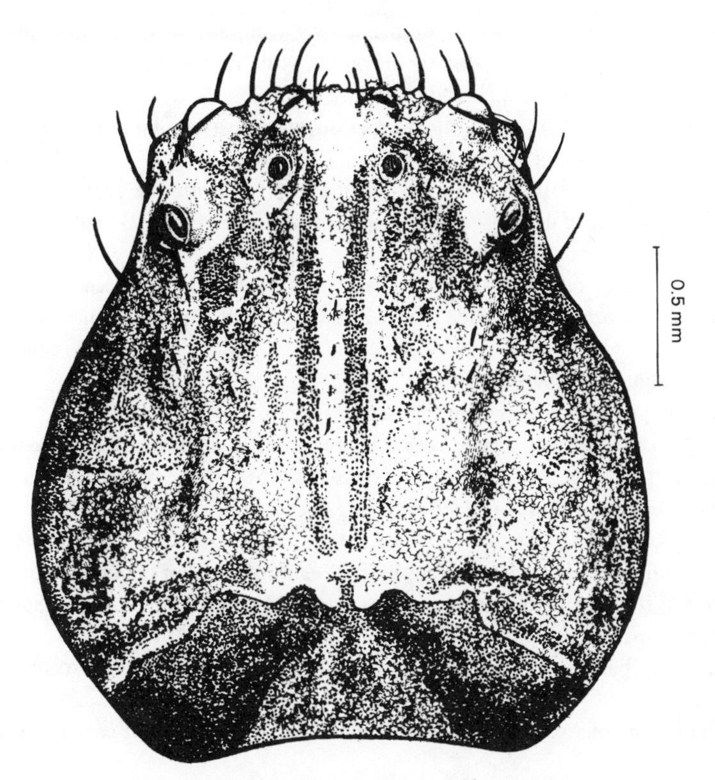

Fig. 108: *Xysticus* C. L. Koch, 1835; female, carapace, dorsal view

Keys to the Species of Xysticus in Israel and Sinai

Males:

1. Body of spider invested with short, strong, blunt (not clavate) bristles — 2
– Body of spiders covered with long, fine, pointed bristles — 7
2. Tibia of palpus with two apophyses (Figs. 127, 128). Tegulum of palpus bears large apophysis projecting towards the tibia (Fig. 127). — **X. bliteus** (Simon)
– Tibia of palpus with three apophyses. Tegulum of palpus bears other structures — 3
3. Ventral apophysis of tibia large, very broad and bears an external protrusion close to the base (Figs. 109, 114) — 4
– Ventral apophysis of tibia finger-like, inclined inwards and lacking external protrusions — 5
4. Tegulum of palpus with a semicircular black ridge backed by two tiny black arches, placed off-centre, close to the central swelling giving rise to the encircling embolar duct (Fig. 109). Adults found from October to December. — **X. rectilineus** (O. P.-Cambridge)
– Tegulum of palpus bears, close to the central swelling giving rise to the encircling embolar duct, a stout black protrusion lacking additional protuberances (Fig. 114). Adults found mainly from April to May. — **X. tristrami** (O. P.-Cambridge)

5. Large retrolateral apophysis of tibia narrows gradually towards tip (Figs. 131, 132). Tegulum of palpus bears a slightly elevated, white fold at middle, partly covering the rising point of the encircling embolar duct (Fig. 131). **X.? bufo** (Dufour)
– Large retrolateral apophysis of tibia has a distinct constriction or recess at about middle or two-thirds the height of apophysis (Figs. 122, 124) 6
6. Terminal part of retrolateral apophysis of tibia (following the recess) rounded except for a deep, crescent-shaped notch on apical inner side (Figs. 22). Tutaculum of palpus triangular and rather protruding. **X. caperatoides** Levy
– Terminal part of retrolateral apophysis of tibia (following the recess) straight, blade-shaped on the apical inner side (Fig. 124). Tutaculum cup-like, almost inconspicuous. **X. caperatus** Simon
7. Tibia of palpus with three apophyses 8
– Tibia of palpus with only two apophyses 9
8. Apical part of bulb of palpus, below tip of embolar duct, bears a hook-like sclerotic apophysis (Figs. 166, 167). A fine, crescent-shaped ridge projects from tegulum of palpus, close to central sclerotization giving rise to embolar duct (Fig. 166). **X. lalandei** (Audouin)
– Bulb of palpus lacking apical apophysis. Tegulum with an encircling, rounded, non-sclerotized swelling, ending in an indistinct opening (Fig. 150). **X. kempeleni** Thorell
9. Tegulum of palpus bears large, distinct apophyses (Figs. 133, 138, 146) 10
– Tegulum of palpus without distinct apophyses, at most with crescent-shaped protrusion 12
10. Tegulum of palpus with T-shaped, hammer-headed apophysis at centre (Figs. 133, 138) 11
– Tegulum of palpus with two large, thick, finger-like, curved apophyses which bend towards each other (Fig. 146). **X. kochi** Thorell
11. Shaft and mesal crossbar of tegular T-shaped apophysis markedly bent (Fig. 133); leaf-like, sclerotized process rises at base of tegulum. Tip of embolus forms large loops (Figs. 133, 134). **X. cristatus** (Clerck)
– Shaft and mesal crossbar of tegular T-shaped apophysis straight (Fig. 138); large cone-like apophysis rises at base of tegulum. Tip of embolus slightly recurved (Figs. 138, 139). **X. edax** (O. P.-Cambridge)
12. Tegulum slightly off-centre, with a sclerotized, protruding, crescent-shaped structure (Fig. 154); tutaculum cup-like and rather conspicuous (Fig. 155). **X. xerodermus** Strand
– Tegulum smooth, without protruding structures; tutaculum large, pointed and strongly projecting (Fig. 164). **X. jezequeli** Karol

Females :

1. Body of spider invested with short, strong, blunt (not clavate) bristles 2
– Body of spider covered with long, fine, pointed bristles 6
2. Epigynum with deep violin-like depression surrounded by dark, heavily sclerotized rims (Fig. 129). **X. bliteus** (Simon)
– Epigynum without such a deep central depression 3
3. Epigynum with sclerotized lateral caps, sometimes in form of arched folds proximally united in basal part of epigynum to form central hood-like structure 4

- – Epigynum without lateral sclerotic caps, but with membranous, twisted folds extending along each side of central hooded structure (Fig. 125). **X. caperatus** Simon
4. Central, partly membranous, hood-like structure extends about half the length of epigynal plate (Fig. 116); curtain-like folds surround opening of the hooded structure. Area close to epigynal furrow covered by two thick platelets bordered by dark rims (Fig. 116). **X. tristrami** (O. P.-Cambridge)
- – Sclerotic, rounded or almost flattened hood-like structure extends less than one-third the length of epigynal plate (Figs. 111, 118). Central part of epigynum usually bears dark, winding bands 5
5. Two large coils, formed by dark, broad bands, lie across most of the central part of the epigynal plate (Fig. 111); sometimes sclerotic bands only partly visible (Fig. 112). Spermathecae form large, closely set, compact and almost rounded bodies (Fig. 113, require dissection). **X. rectilineus** (O. P.-Cambridge)
- – Two small, rounded bodies, formed of dark, strongly coiled bands, lie below inner margins of the large, lateral caps located on each side of epigynal plate (Figs. 118, 119). Spermathecae form large, partly lobed bodies (Fig. 20; require dissection). **X. ferus** O. P.-Cambridge
6. Epigynum with central hood-like structure (Figs. 158, 168) 7
- – Epigynum without central hood-like structure 8
7. Central hood-like structure of epigynum in form of elongated median protrusion extending to about two-thirds the length of epigynal plate (Fig. 158); hood-like structure accompanied on each side by large, twisted folds. **X. graecus** C. L. Koch
- – Hood-like structure of epigynum small, with transparent walls and triangular, tent-like opening (Fig. 168); sclerotic lateral caps located on each side of hood-like structure. **X. lalandei** (Audouin)
8. Epigynum in form of raised, convex shield with two dark, rimmed orifices on both sides (Fig. 135); area below orifices sometimes marked by two dark, convergent, narrow furrows (Fig. 136). **X. cristatus** (Clerck)
- – Epigynal plate shaped otherwise 9
9. Epigynum divided by distinct median septum (Fig. 148). **X. kochi** Thorell
- – Epigynum without distinct median septum 10
10. Spermathecae forming large, compact, almost rounded bodies (Figs. 161, 163; require dissection) 11
- – Spermathecae in form of thick, convolute, partly constricted tubes (Figs. 144, 153, 157; require dissection) 12
11. Central part of epigynal plate slants gradually towards epigastric furrow; sides of central plate marked by dark, S-shaped rims (Fig. 162). **X. gymnocephalus** Strand
- – Central part of epigynal plate in form of rounded depression bordered by distinct, raised rim located on side close to the epigastric furrow (Fig. 160). **X. altitudinis** Levy
12. Spermathecae accompanied by partly sclerotized, concave platelets (Figs. 144, 145; require dissection). **X. edax** (O. P.-Cambridge)
- – Spermathecae without additional platelets 13
13. Epigynal plate in form of oblong depression, partly surrounded by double set of thick rims (Fig. 152). Tubes of spermathecae on median inner side form large loops lacking intermediary constrictions (Fig. 153; require dissection). **X. kempeleni** Thorell
- – Epigynal plate marked by thick sclerotic walls partly surrounding rounded central

depression with an opening leading towards the epigastric furrow (Fig. 156); septum-like elevation extends half-way into central depression through its opening. Spermathecae consist of thick, markedly constricted tubes not extending medially into long, free loops (Fig. 157; requires dissection). **X. xerodermus** Strand

Section A. Species with Short, Strong and Blunt Bristles

Xysticus rectilineus (O. P.-Cambridge, 1872)
Figs. 109–113

Thomisus rectilineus O. P.-Cambridge, 1872, *Proc. zool. Soc. Lond.*, p. 306.
Xysticus tristrami (O. P.-Cambridge, 1872). Pavesi, 1895, *Boll. Musei Zool. Anat. Comp. R. Univ. Torino*, 10 : 8 (misdetermination).
Xysticus rectilineus (O. P.-Cambridge, 1872). Kulczyński, 1911, *Bull. Acad. Sci. Cracovie*, p. 31, figs. 30, 32, 33; Roewer, 1954, *Katalog der Araneae*, 2 (1): 906; Bonnet, 1959, *Bibliographia Araneorum*, 2 (5): 4896; Levy, 1976, *Israel J. Zool.*, 25: 4.

Length of male 4.2–5.1 mm, female 5.1–9.4 mm. Coloration of dorsal side of prosoma brown to dark red in males, deep yellowish-brown in females. Opisthosoma of male white with dark, broad, transverse stripes; females only with light-coloured bands across posterior part.
Male Palpus: Tibia with ventral, intermediate and retrolateral apophyses (Figs. 109, 110). Ventral apophysis broad, curved inwards; shaft of apophysis long, oblique, on mesal side with upper corner mesally protruding; outer side much shorter, deeply notched at middle, with downwards projecting extensions (Fig. 109). Intermediate apophysis thumb-like, round and stout (Fig. 110). Retrolateral apophysis long, thick, pointed and slightly inclined. Transparent tutaculum cup-like, almost inconspicuous (Fig. 110). Tegulum with large, black, round swelling at centre, giving rise to encircling embolar duct (Fig. 109); close to central swelling, off-centre, rises a black, semicircular tegular apophysis, backed by two tiny black arches (Fig. 109). Tip of embolus bends slightly at tutacular edge (Fig. 110).
Female Epigynum: Epigynal plate with slightly variable sclerotizations (Figs. 111, 112). Basal part of plate with a wide or rounded, hood-like structure with an opening directed towards epigastric furrow; pair of dark, slender arched caps, occasionally rather broad, rising on each side of corners of hood with their openings directed towards each other (Figs. 111, 112). Central part of epigynal plate usually with dark, winding folds (Fig. 111) or folds less conspicuously marked or almost indistinct (Fig. 112). Spermathecae are usually without structural variation (Fig. 113).

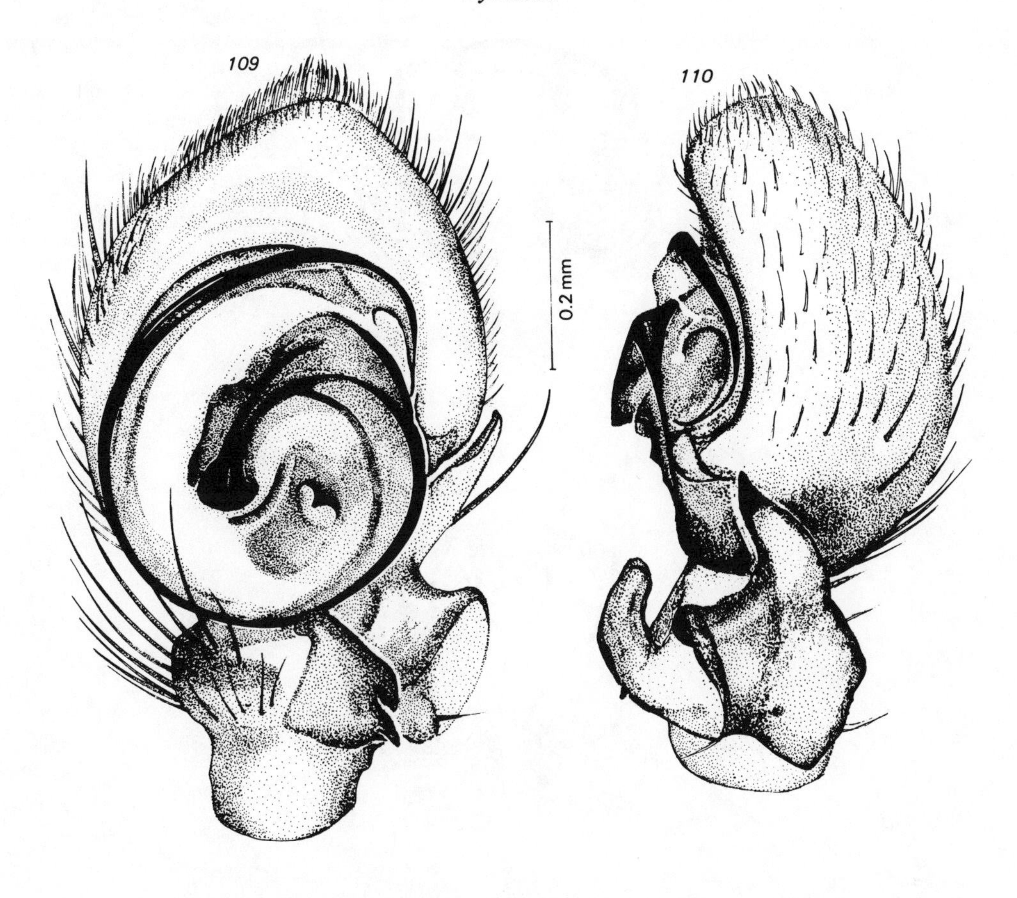

Figs. 109–110: *Xysticus rectilineus* (O. P.-Cambridge, 1872); male, left palpus
109. ventral view; 110. lateral view

Distribution: Syria, Lebanon, Israel.
Israel: From Mt. Hermon (19) throughout the country to Lahav (15), 'Arad and Yeroẖam (17).
It is the most common of the *Xysticus* species of Israel. Adults of the two sexes are found from October to December; only adult females are usually found under stones thereafter, from January to May.

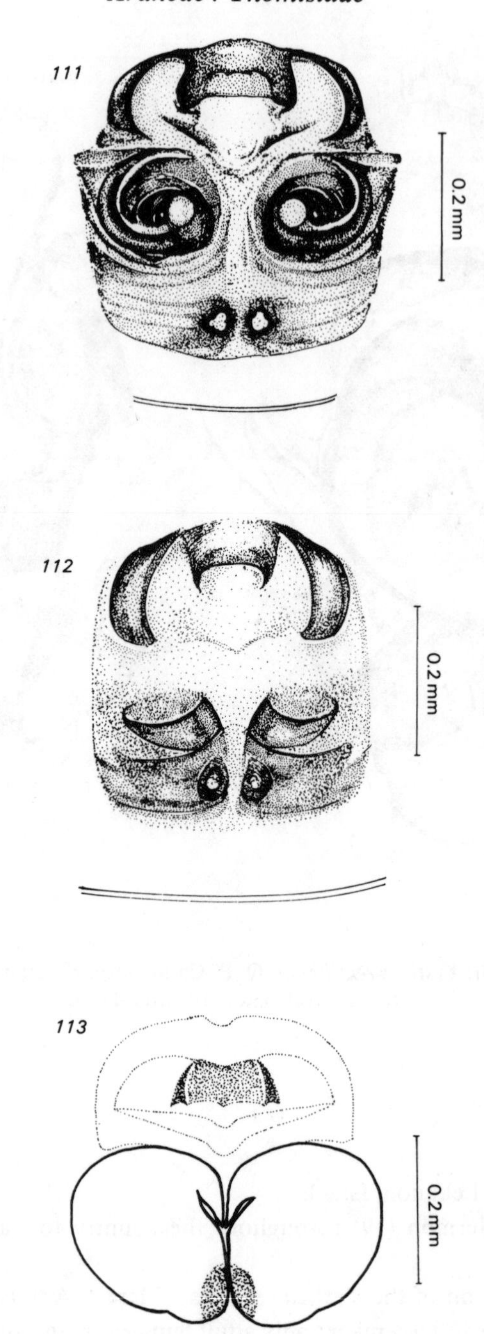

Figs. 111–113: *Xysticus rectilineus* (O. P.-Cambridge, 1872); female
111. epigynum, common form; 112. epigynum, variation; 113. spermathecae, inner view

Xysticus tristrami (O. P.-Cambridge, 1872)
Figs. 114–117

Thomisus tristrami O. P.-Cambridge, 1872, *Proc. zool. Soc. Lond.*, p. 304, pl. 14, fig. 16.
Xysticus tristrami —. Simon, 1884, *Annls Soc. ent. Fr.*, (6) 4: 185; Kulczyński, 1911, *Bull. Acad.* Sci. Cracovie, pp. 32–35, figs. 31, 34, 35; Roewer, 1954, *Katalog der Araneae*, 2 (1): 910; Levy, 1976, *Israel J. Zool.*, 25: 7.

Length of male 3.6–4.5 mm, female 5.4–8.0 mm. Coloration of prosoma and legs deep reddish-brown, sometimes mottled with white. Opisthosoma grey to yellowish, with white, brown or black markings.
Male Palpus: Tibia with ventral, intermediate and retrolateral apophyses (Figs. 114, 115). Ventral apophysis oblique, markedly curved inwards, with round swelling at apex projecting mesally (Fig. 114) and with small cone-like protrusion at base pointing outwards (Figs. 114, 115). Intermediate apophysis large, oval, leaf-like in ventral

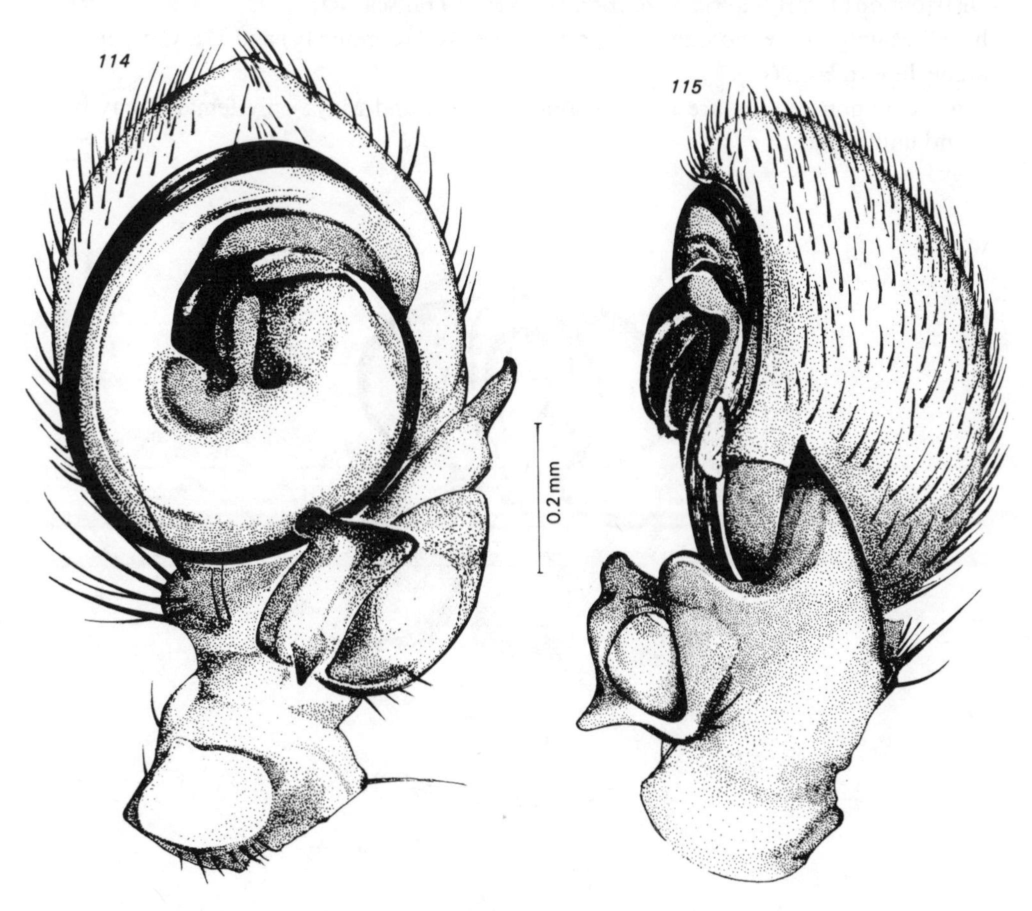

Figs. 114–115: *Xysticus tristrami* (O. P.-Cambridge, 1872); male, left palpus
114. ventral view; 115. lateral view

view (Fig. 114), with very wide base and round top in lateral view (Fig. 115). Retrolateral apophysis wide at base and with dark, sclerotic, slightly bent tip (Figs. 114, 115). Transparent tutaculum inconspicuous, rather large, inclined outwards (Fig. 115). Tegulum with black sclerotic protuberance in central to upper part, giving rise to encircling embolar duct (Fig. 114); tegular apophysis of medium size rises close to central protrusion (Fig. 114). Straight embolar tip ends at tutaculum (Fig. 115).

Female Epigynum: Epigynal plate with central, partly membranous, hood-like structure accompanied on each side by two large, sclerotic lateral caps (Fig. 116); opening of central hood with curtain-like folds and occasionally deeply notched at top; distal margins of lateral caps thickened and bending outwards (Fig. 116). Distal part of epigynal plate with an elevated platelet on each side bordered by a dark band (Fig. 116; dorsal-inner view, Fig. 117). Sclerotization of various parts of epigynal plate occasionally varies.

Distribution: Israel, ? Syria, ? southern Turkey, ? Transcaspia.

Israel: Mainly in the mountainous parts, from Mt. Hermon (19) and Mt. Carmel (3) to the Judean Hills (11).

Quite common; adults are found mainly in April and May; only females may be found until August.

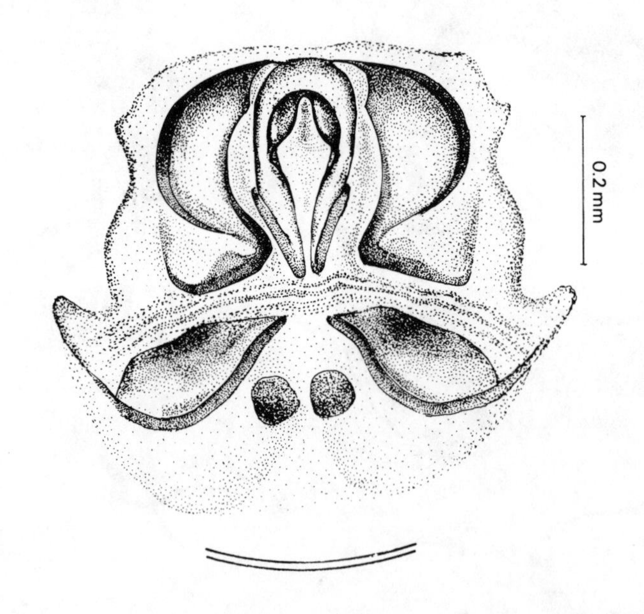

Figs. 116: *Xysticus tristrami* (O. P.-Cambridge, 1872); female, epigynum

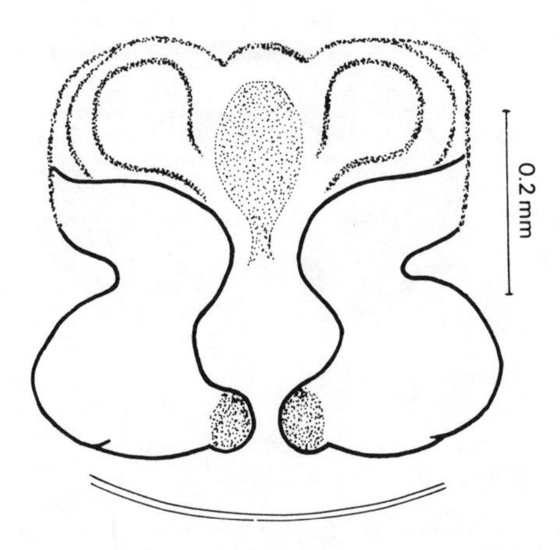

Fig. 117: *Xysticus tristrami* (O. P.-Cambridge, 1872); female, spermathecae, inner view

Xysticus ferus O. P.-Cambridge, 1876
Figs. 118–120

Xysticus ferus O. P.-Cambridge, 1876, *Proc. zool. Soc. Lond.*, p. 583; Roewer, 1954, *Katalog der Araneae*, 2 (1): 901; Bonnet, 1959, *Bibliographia Araneorum*, 2 (5): 4875; Levy, 1976, *Israel J. Zool.*, 25: 9.

Xysticus tristrami (O. P.-Cambridge, 1872). Pavesi, 1895, *Boll. Musei Zool. Anat. Comp. R. Univ. Torino*, 10 : 8 (misdetermination); Kulczyński, 1908, *Bull. Acad. Sci. Carcovie*, pp. 73–75, pl. 2, fig. 14 (misdetermination).

Xysticus similis Kulczyński, 1911, *Bull. Acad. Sci. Cracovie*, p. 33.

Male unknown. Length of female 6.2–8.1 mm. Coloration of prosoma yellowish anteriorly and deep brown posteriorly. Opisthosoma dull yellowish-brown traversed by light bands.

Female Epigynum: Epigynal plate slightly variable (Figs. 118, 119). Basal part of plate with low, almost flat, central hood-like structure, with large, sclerotized, lateral caps on each side (Fig. 118); occasionally, lateral caps in form of arched folds united basally with the central hood (Fig. 119). Central part of epigynum with dark, winding bands. Distal part of epigynal plate with two triangular platelets (Fig. 118), the latter occasionally indistinct (Fig. 119). On dorsal-inner side of epigynum, the basal winding of the folds is best discerned in less sclerotized epigynal plates (Fig. 120).

Distribution: Egypt, Sinai, Israel, ?Jordan.
Israel : Judean Desert (12) and Dead Sea area (13).
Sinai : Wadi Yaḫ'med, southern Sinai (22).
Adult females are found mainly from December to April ; a few also occur in May to July.

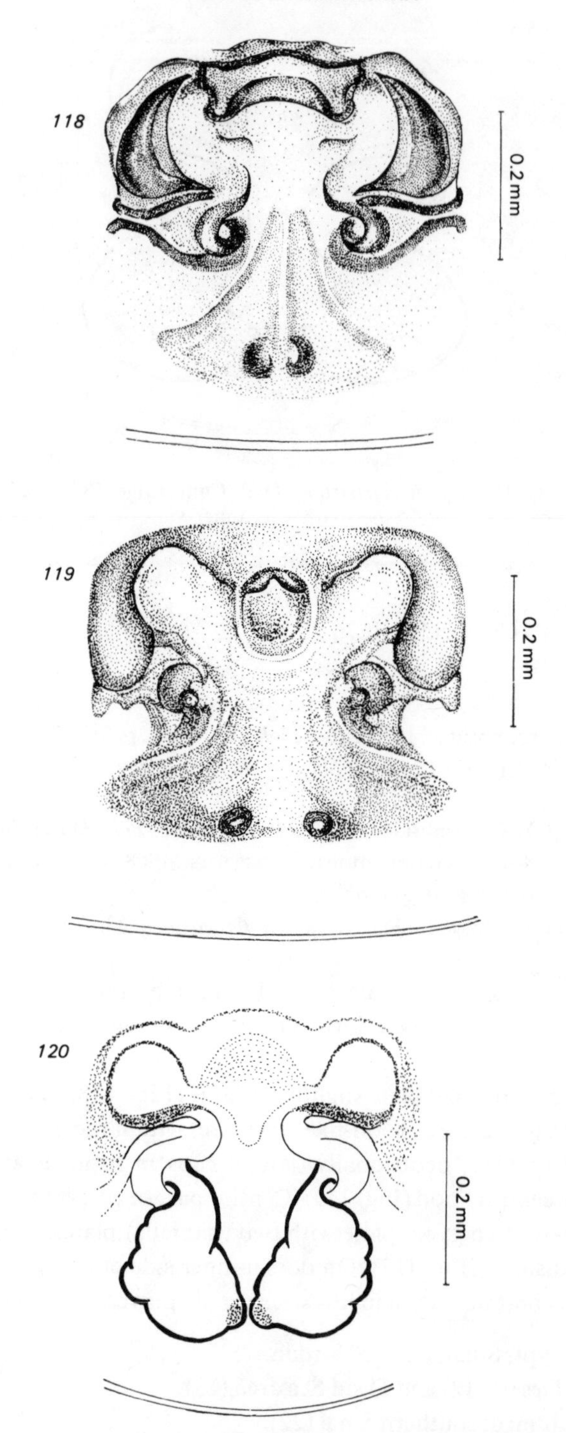

Figs. 118–120: *Xysticus ferus* O. P.-Cambridge, 1876; female
118. epigynum, common form; 119. epigynum, variation; 120. spermathecae, inner view

Xysticus caperatoides Levy, 1976
Figs. 121, 122

Xysticus caperatoides Levy, 1976, *Israel J. Zool.*, 25: 11.

Length of male 3.5–4.0 mm; female unknown. Coloration of prosoma and legs deep reddish-brown; legs also feature fine white stripes. Opisthosoma mottled with small white, brown and black spots.
Male Palpus: Palp relatively small. Tibia with ventral, intermediate and retrolateral apophyses. Ventral apophysis finger-like, slightly inclined mesally and inwards (Figs. 121, 122). Intermediate apophysis thick, apically round, separated from retrolateral apophysis by deep notch (Fig. 122). Retrolateral apophysis constricted at about two-thirds of its height; following part rounded except for a deep recess on upper-inner side (Fig. 122). Tutaculum triangular, conspicuously protruding. Tegulum, at middle, giving rise to dark, sclerotized encircling embolar duct (Fig. 121); a low tegular elevation, finely separated at middle from rising point of the encircling duct, extends towards tutaculum (Fig. 122).

Distribution, Israel: Jerusalem (11) and Dead Sea area (13).
Adult males are found in May–June. The matching female might be represented by *X. ferus.*

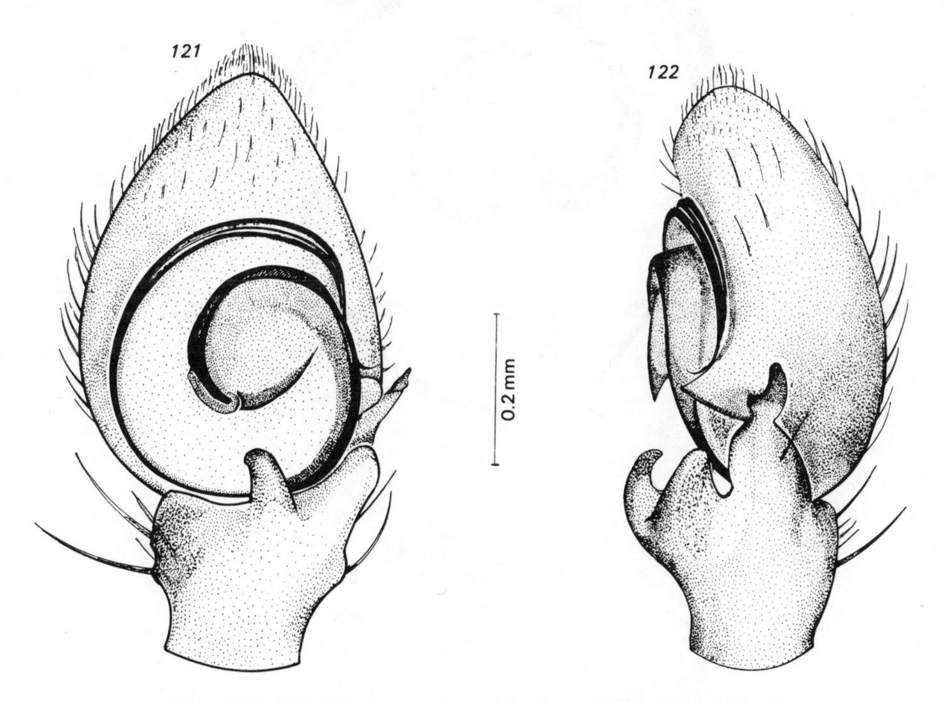

Figs. 121–122: *Xysticus caperatoides* Levy, 1976; male, left palpus
121. ventral view; 122. lateral view

Xysticus caperatus Simon, 1875
Figs. 123–126

Xysticus caperatus Simon, 1875, *Les Arachnides de France*, 2: 198; Roewer, 1954, *Katalog der Araneae*, 2 (1): 898; Bonnet, 1958, *Bibliographia Araneorum*, 2 (4): 3792 (as *Proxysticus*); Levy, 1976, *Israel J. Zool.*, 25: 12.

Length of male 5.1 mm, female 6.6–8.0 mm. Coloration of prosoma and proximal joints of legs deep brown in male, light brown veined with yellow in female. Opisthosoma of male dark brown on back surrounded by white, of female light-coloured with scattered dark brown markings.

Male Palpus: Tibia with ventral, intermediate and retrolateral apophyses. Ventral apophysis inclined slightly inwards; apex with mesal recess (Figs. 123, 124). Intermediate apophysis short, cone-like, widely separated from retrolateral apophysis

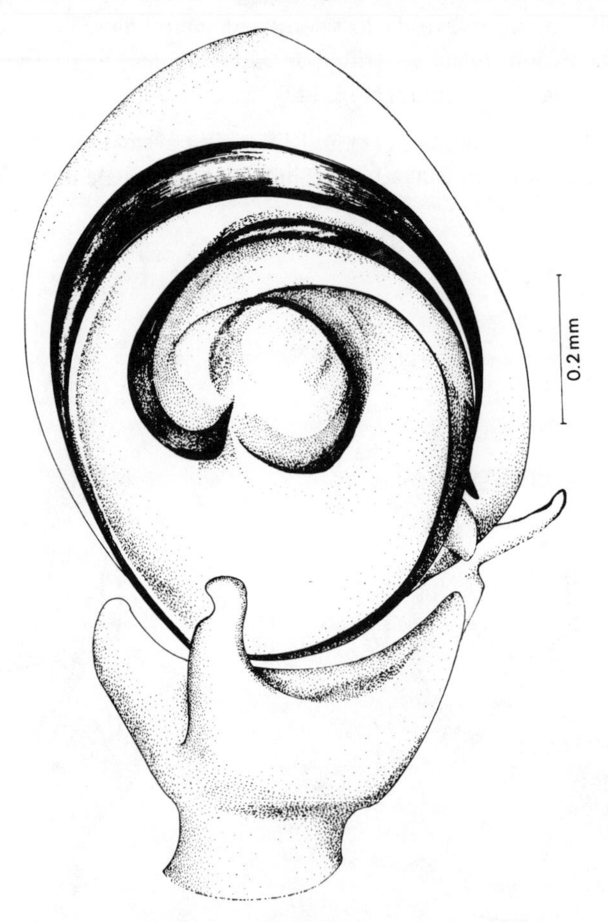

Fig. 123: *Xysticus caperatus* Simon, 1875; male, left palpus, ventral view

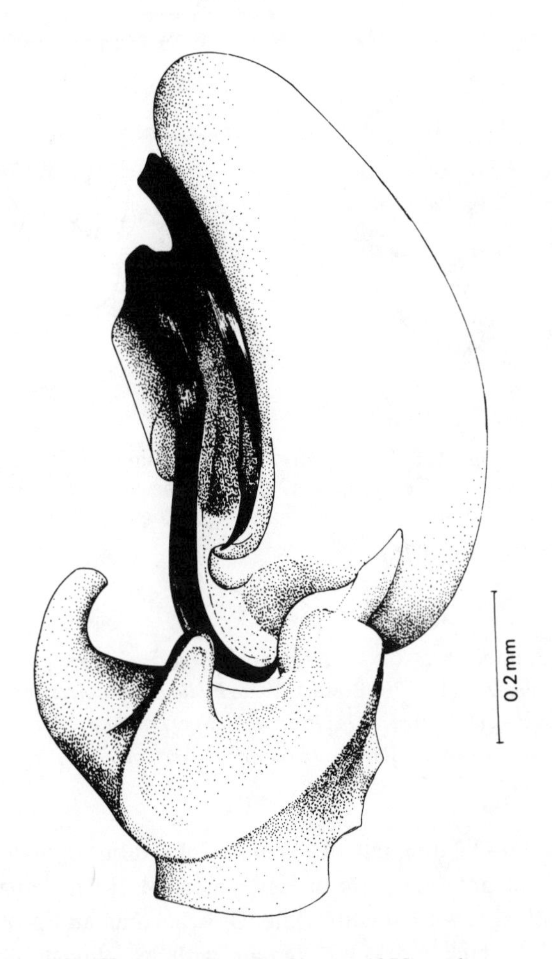

Fig. 124: *Xysticus caperatus* Simon, 1875; male, left palpus, lateral view

(Fig. 124). Retrolateral apophysis with step-like recess on outer side; following part blade-like, straight on upper-inner side (Fig. 124). Tutaculum cup-like, almost inconspicuous (Fig. 124). Tegulum encircled by broad, black, sclerotized, embolar duct (Fig. 123); at centre, dark raised rims surround a low depression (Fig. 123).
Female Epigynum: Epigynal plate from centre to base with membranous hood-like structure seemingly consisting of three joined hoods located one behind the other (Fig. 125). Large, membranous, twisted folds extend along each side of central structure (Fig. 125). Spermathecae on dorsal surface, each with an occasional row of about six peculiar, tiny, tube-like stubs (Fig. 126).

Distribution : Spain, Corsica, Sardinia, Algeria, Tunisia, Israel.
Israel: Golan Heights (18), Mt. Carmel (3) to the Judean Hills (11).
Adults are found from May to July.

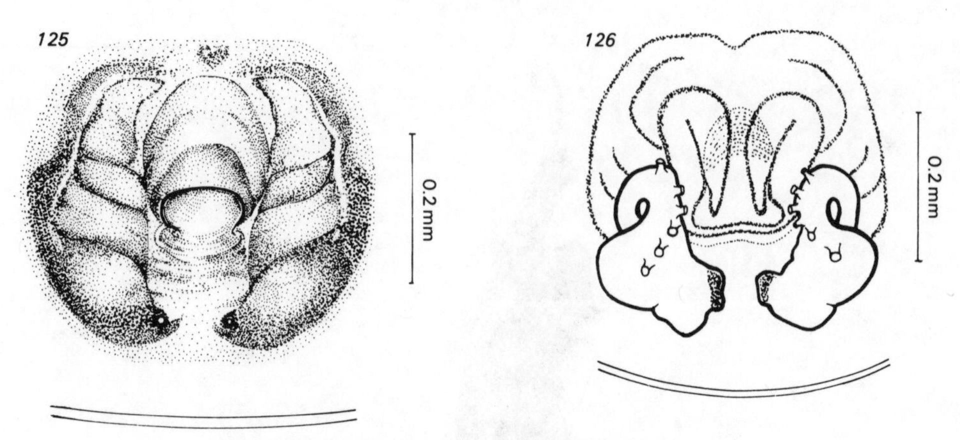

Figs. 125–126: *Xysticus caperatus* Simon, 1875; female
125. epigynum; 126. spermathecae, inner view

Xysticus bliteus (Simon, 1875)
Figs. 127–130

Oxyptila blitea Simon, 1875, *Les Arachnides de France*, 2: 236, pl. 7, fig. 25; Roewer, 1954, *Katalog der Araneae*, 2 (1): 876; Bonnet, 1958, *Bibliographia Araneorum*, 2 (4): 3252.
Xysticus hirtus O. P.-Cambridge, 1876, *Proc. zool. Soc. Lond.*, p. 581.
Xysticus bliteus (Simon, 1875). Pavesi, 1878, *Annali Mus. civ. Stor. Nat. Genova*, 11: 351; Levy, 1976, *Israel J. Zool.*, 25: 15.

Length of male 3.0–3.3 mm, female 5.3 mm. Coloration: in male, entire prosoma and proximal joints of anterior pairs of legs almost black; in female, posterior part of prosoma and all legs mottled with white. Opisthosoma and posterior pairs of legs of male white with black markings; female with an almost black back, but light-coloured on sides and venter.
Male Palpus: Tibia with ventral and retrolateral apophyses. Ventral apophysis large, laterally positioned and inclined obliquely outwards (Fig. 127); apex with mesally projecting round swelling and with small protuberance on outer corner (Fig. 127). Retrolateral apophysis projecting markedly outwards (Fig. 127), with S-like stem in lateral view (Fig. 128). Tegulum encircled by broad, black, sclerotic, embolar duct (Figs. 127, 128) and with large tegular apophysis projecting obliquely downwards with slightly notched tip (Fig. 127).
Female Epigynum: Epigynal plate with deep violin-like depression, traversed longitudinally by slight elevation (Fig. 129); except for light-coloured membranous fold at basal part, entire area surrounding central depression dark, heavily sclerotized and traversed by numerous folds (Fig. 129; dorsal-inner view, Fig. 130).

Distribution: Mediterranean countries.
Israel: Coastal Plain (8, 9), Valley of Yizre'el (5), Judean Hills (11), ? Jericho (13).
Adults are found from September to November.

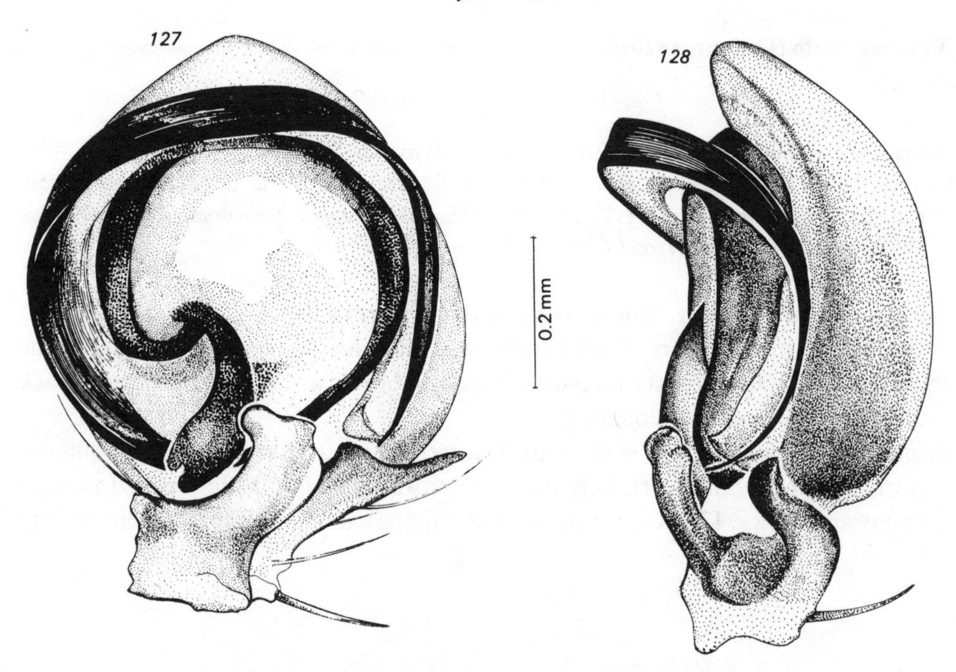

Figs. 127–128: *Xysticus bliteus* (Simon, 1875); male, left palpus
127. ventral view; 128. lateral view

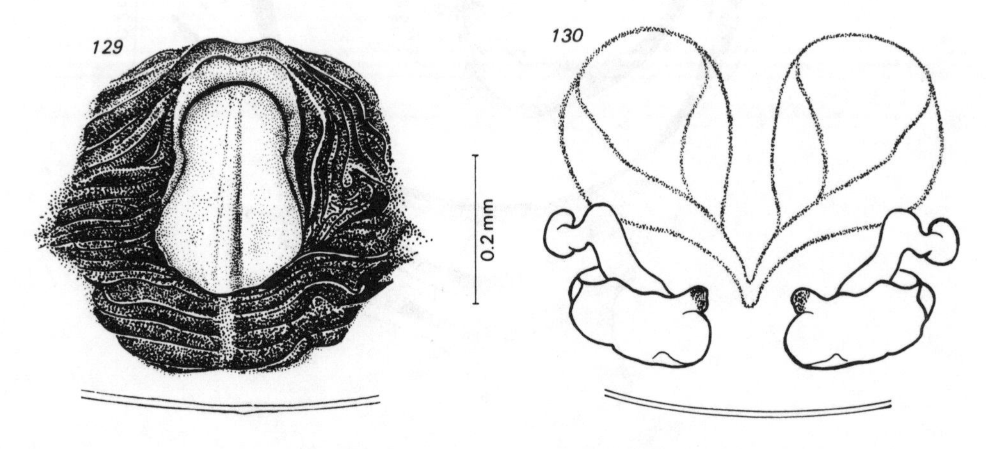

Figs. 129–130: *Xysticus bliteus* (Simon, 1875); female
129. epigynum; 130. spermathecae, inner view

Xysticus ? bufo (Dufour, 1820)
Figs. 131, 132

Thomiseus bufo Dufour, 1820, *Ann. Gén. Sci. Phys., Bruxelles*, 5: 206.
Xysticus bufo —. Simon, 1864, *Histoire Naturelle des Araignées*, Paris, p. 427; Roewer, 1954, *Katalog der Araneae*, 2 (1): 898; Bonnet, 1958, *Bibliographia Araneorum*, 2 (4): 3790 (as *Proxysticus*); Levy, 1976, *Israel J. Zool.*, 25: 17.

Length of male 4.8 mm; female as yet unknown from Israel. Coloration of male: prosoma and anterior pairs of legs deep brown, with a fine white line in front across the eyes. Opisthosoma and posterior pairs of legs light brown with white markings on back and sides, but venter light-coloured.

Male Palpus: Large. Tibia with ventral, intermediate and retrolateral apophyses. Ventral apophysis almost entirely directed outwards and with tip inclined mesally and inwards (Figs. 131, 132). Intermediate apophysis short, slightly pointed and

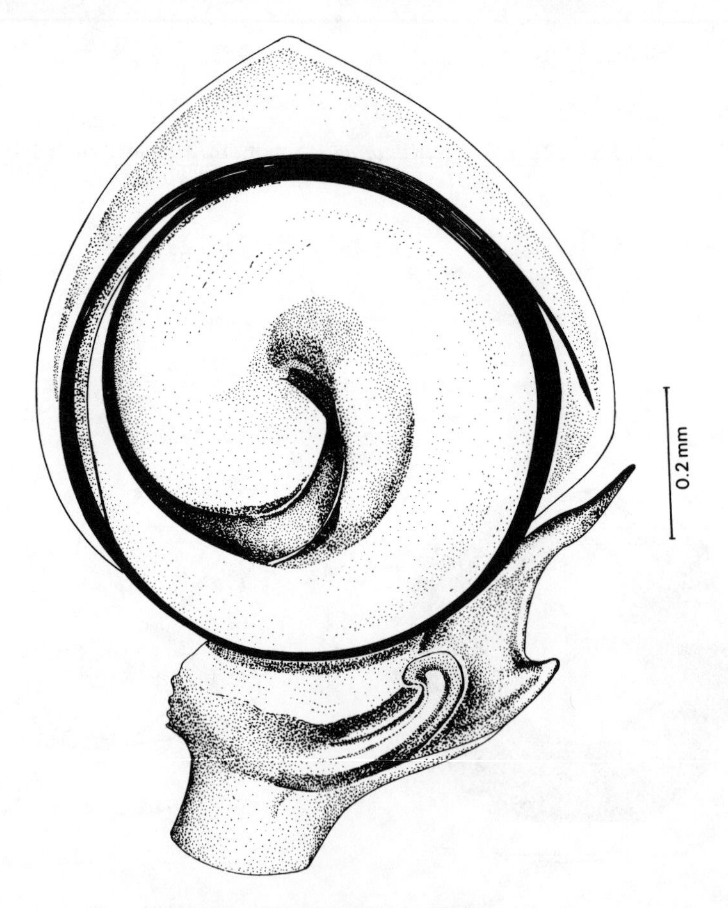

Fig. 131: *Xysticus ? bufo* (Dufour, 1820); male, left palpus, ventral view

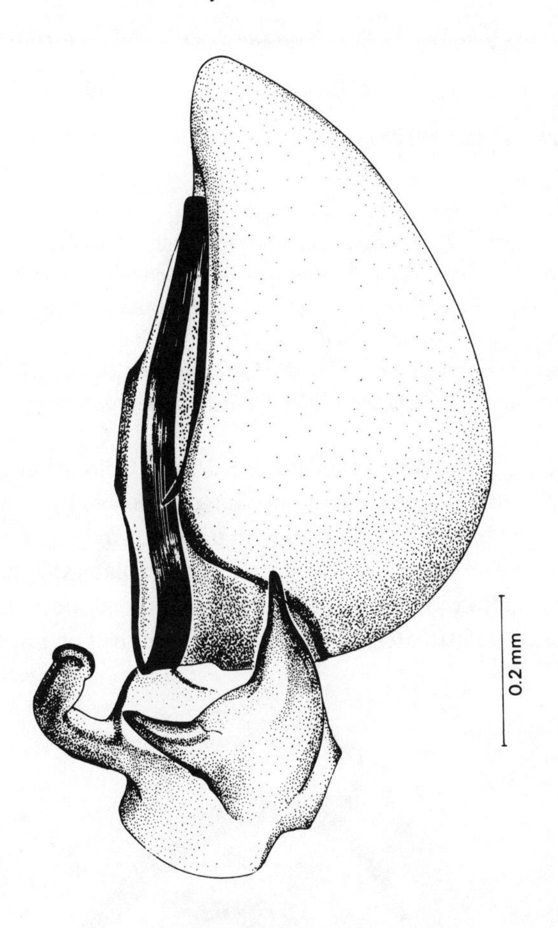

Fig. 132: *Xysticus ? bufo* (Dufour, 1820); male, left palpus, lateral view

broadly joined to retrolateral apophysis (Figs. 131, 132). Retrolateral apophysis large, wide at base and with slightly pointed tip (Figs. 131, 132). Tegulum with slightly elevated white fold at middle, partly covering rising point of encircling dark embolar duct (Figs. 131, 132).

Distribution: Mediterranean countries.
Israel: Southern Coastal Plain (9).
A young male was collected in May; in the laboratory it moulted to adult in July. Material available in European collections is unsatisfactory for the final definition of this species.

Section B. Species with Long, Fine and Pointed Bristles

Xysticus cristatus (Clerck, 1757)
Figs. 133–137

Araneus cristatus Clerck, 1757, *Aranei Suecici*, Stockholm, p. 136.
Xysticus cristatus —. Thorell, 1872, *Remarks on Synonyms of European Spiders*, Upsala, p. 236; Roewer, 1954, *Katalog der Araneae*, 2 (1) : 899; Bonnet, 1959, *Bibliographia Araneorum*, 2 (5) : 4863; Levy, 1976, *Israel J. Zool.*, 25 : 18.
Xysticus promiscuus O. P.-Cambridge, 1876, *Proc. zool. Soc. Lond.*, p. 581; Roewer, 1954, *Katalog der Araneae*, 2 (1) : 906; Bonnet, 1959, *Bibliographia Araneorum*, 2 (5) : 4894.

Length of male 3.3–4.2 mm, female 5.2–8.2 mm. Coloration of prosoma reddish-brown with a wedge-shaped dorsal marking enclosed by a whitish-yellow band. Opisthosoma deep brown on sides and with pale brown dentate pattern on back.
Male Palpus: Tibia short with stout ventral and retrolateral apophyses. Ventral apophysis almost quadrate; outer side slightly rounded and inner side with deep concavity (Figs. 133, 134). Retrolateral apophysis low, rounded on inner side (Fig. 133); top

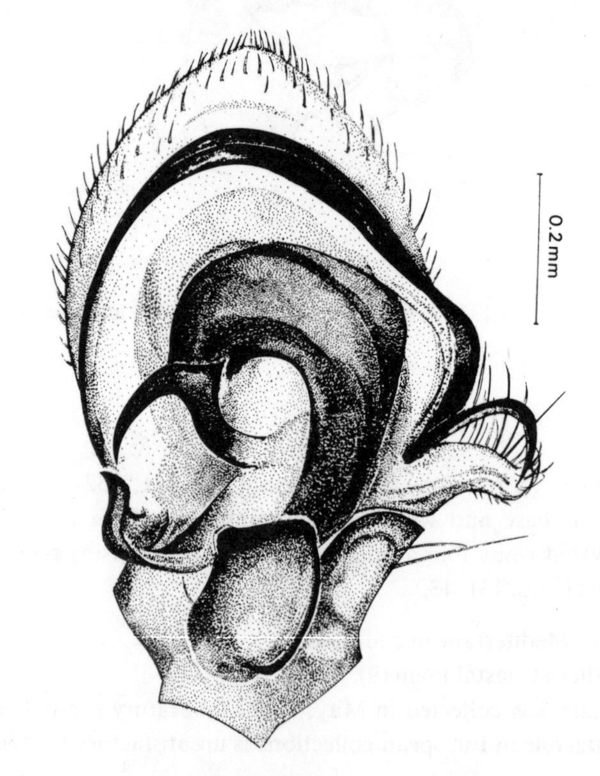

Fig. 133 : *Xysticus cristatus* (Clerck, 1757); male, left palpus, lateral view

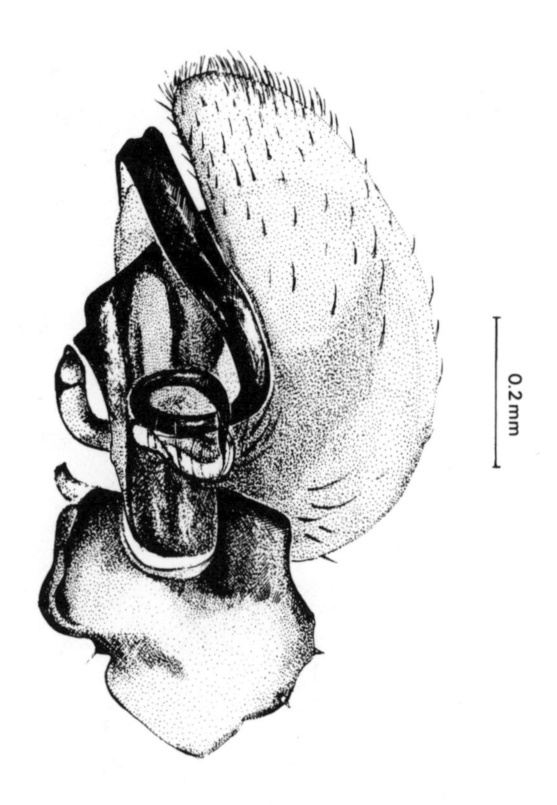

Fig. 134: *Xysticus cristatus* (Clerck, 1757); male, left palpus, lateral view

flattened and slightly raised mesally (Fig. 134). Tutaculum rather large, slightly winding and markedly projecting outwards (Fig. 133). Tegulum with large T-shaped apophysis at centre, its shaft markedly bent; outer crossbar very short and pointed, mesal crossbar long, tapering and inclined downwards (Fig. 133). Tegulum on lower margin with large, leaf-like, sclerotized process, with an almost inconspicuous small spine on inner side (Fig. 133). Tip of embolus loop-like (Figs. 133, 134).
Female Epigynum: Epigynal plate shield-shaped, light-coloured, markedly convex; spermathecal orifices enclosed by dark sclerotic rims opening obliquely on both sides of basal part of epigynal plate (Fig. 135); occasionally, two dark narrow furrows originate from the spermathecal orifices, curving outwards and obliquely backwards, extending distally towards epigastric furrow (Fig. 136; dorsal-inner view, Fig. 137).

Distribution: Almost throughout the Palaearctic region.
Israel: Along the Coastal Plain (4, 8, 9).
Adults are found in March–April.

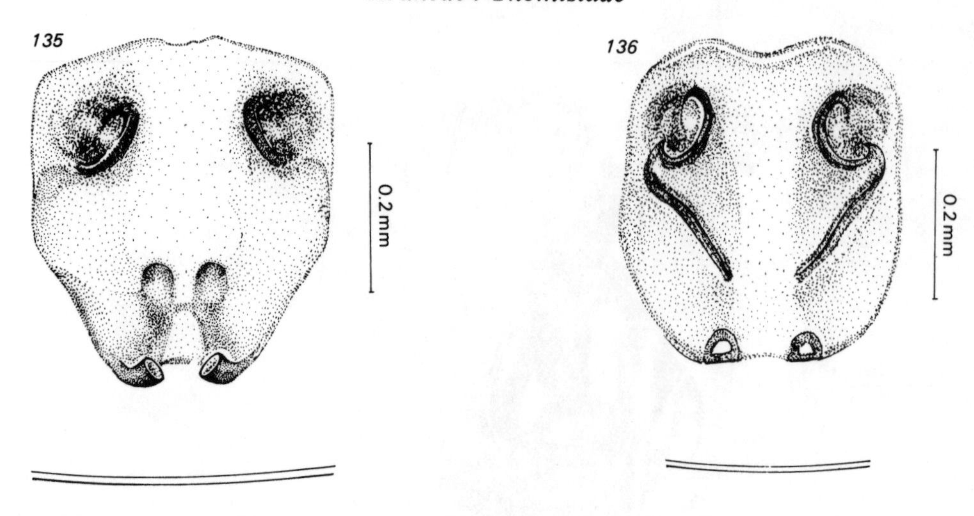

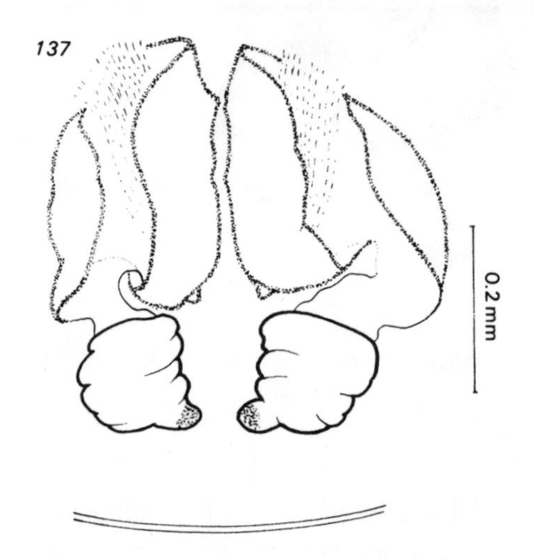

Figs. 135–137: *Xysticus cristatus* (Clerck, 1757); female
135. epigynum, common form; 136. epigynum, variation; 137. spermathecae, inner view

Xysticus edax (O. P.-Cambridge, 1872)
Figs. 138–145

Thomisus edax O. P.-Cambridge, 1872, *Proc. zool. Soc. Lond.*, p. 303.

Xysticus cristatus (Clerck, 1757). Pavesi, 1895, *Boll. Musei Zool. Anat. Comp. R. Univ. Torino*, 10: 8 (misdetermination).

Xysticus edax (O. P.-Cambridge, 1872). Reimoser, 1919, *Abh. zool.-bot. Ges. Wien*, 10 (2): 128; Roewer, 1954, *Katalog der Araneae*, 2 (1): 900; Bonnet, 1959, *Bibliographia Araneorum*, 2 (5): 4870; Levy, 1976, *Israel J. Zool.*, 25: 21.

Length of male 2.9–3.6 mm, female 5.2–7.5 mm. Coloration of sides of prosoma and legs brown to dark brown mottled with white. Back of prosoma with a wedge-shaped yellowish marking. Opisthosoma with a median light-coloured dentate band on back, and with brown sides.

Male Palpus: Ventral and retrolateral apophyses of tibia short and stout. Ventral apophysis almost flattened on top, slightly protruding mesally (Fig. 138). Retrolateral apophysis almost quadrate in lateral view (Fig. 139). Tutaculum forming long, slender process (Fig. 139). Tegulum, at centre, with T-shaped or hammer-headed apophysis with short, straight shaft (Fig. 138); outer crossbar with rounded tip pointing obliquely upwards; mesal crossbar much longer, straight, tapering and extending over bulb's edges (Fig. 138). Tegulum on lower part with large, dark, cone-like apophysis (Fig. 138). Tip of embolus slightly recurved.

Female Epigynum: Epigynal plate comprises a small central orifice, completely or partly encircled by a dark, sclerotic band (Figs. 140–143); occasionally, surrounding band folds slightly inwards on basal edge of central orifice (Figs. 140, 141). Shallow, canal-like depression extends from bottom of central orifice towards epigastric furrow. Accompanying structures of spermathecae vary considerably with regard to grades of sclerotization, but basic plan remains identical (Figs. 144, 145).

Distribution, Israel: Throughout the country from Mt. Hermon (19), the Coastal Plain (8) and Judean Desert (12) to the Northern Negev (15).

Adults are found under stones or on the ground among low grass from about February to May.

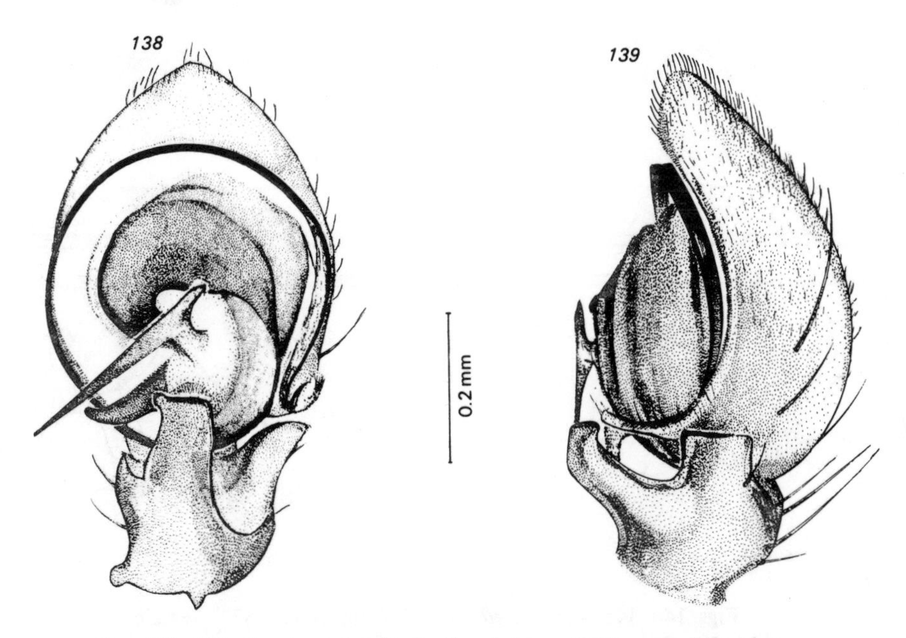

Figs. 138–139: *Xysticus edax* (O. P.-Cambridge, 1872); male, left palpus
138. ventral view; 139. lateral view

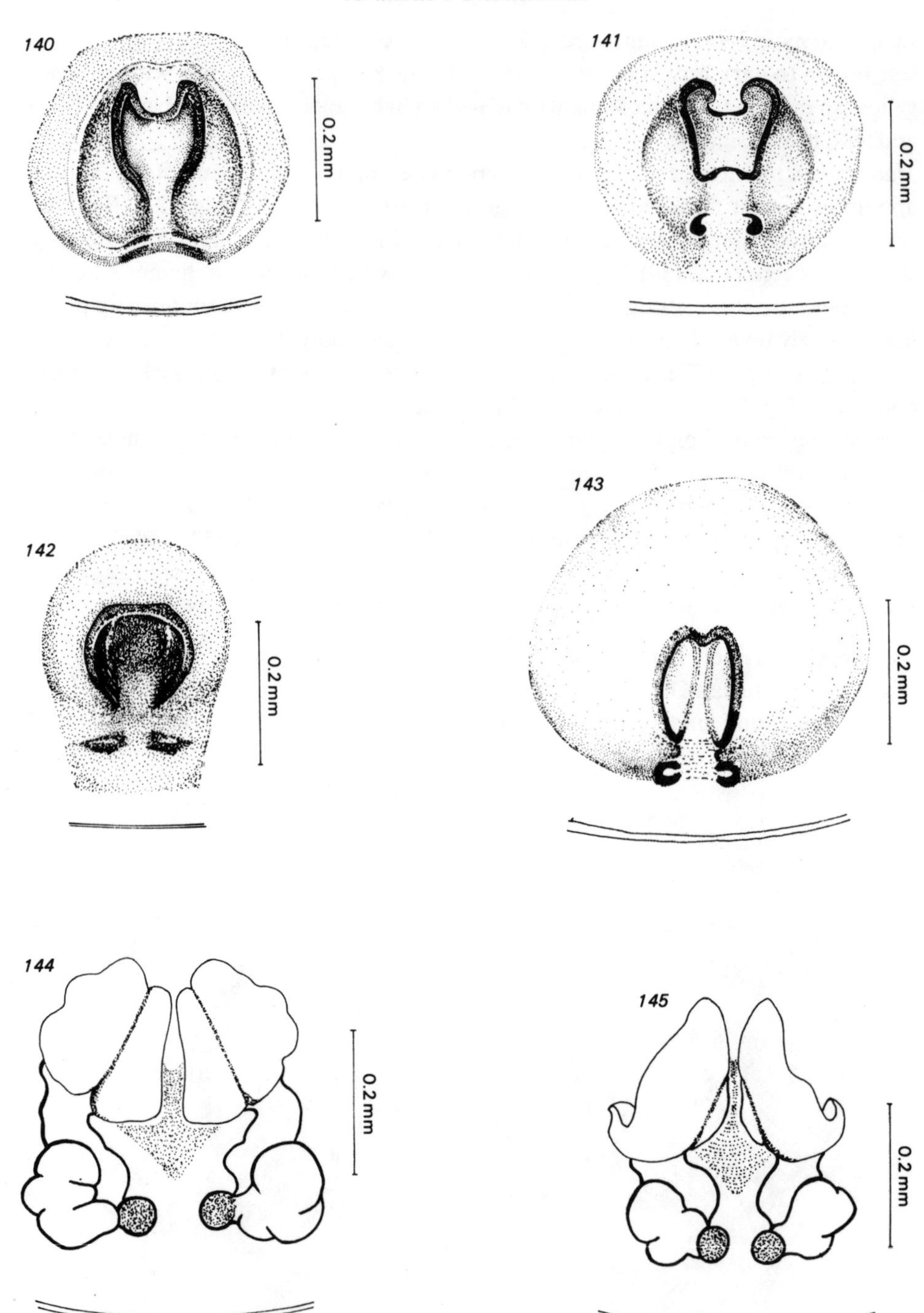

Figs. 140–145: *Xysticus edax* (O. P.-Cambridge, 1872); female
140. epigynum, common form; 141–143. epigynum, variations;
144. spermathecae, common form, inner view; 145. spermathecae, variation, inner view

Xysticus kochi Thorell, 1872
Figs. 146–149

Xysticus viaticus C. L. Koch, 1835, Arachniden, in: G. W. F. Panzer, *Faunae Insectorum Germaniae initia*, Regensburg, Heft 130, pl. 13, 14 (preoccupied).
Xysticus kochi Thorell, 1872, *Remarks on Synonyms of European Spiders*, Upsala, p. 241; Roewer, 1954, *Katalog der Araneae*, 2 (1): 902; Bonnet, 1959, *Bibliographia Araneorum*, 2 (5): 4880; Levy, 1976, *Israel J. Zool.*, 25: 23.

Length of male 4.6 mm, female 8.4 mm. Coloration of prosoma deep brown with a wedge-shaped brown marking on back enclosed by a yellowish band. Opisthosoma with a median, pale brown, dentate pattern on back and with brown sides.
Male Palpus: Tibia short with stout ventral and retrolateral apophyses (Figs. 146, 147). Ventral apophysis pointed apically and slightly notched mesally (Fig. 146).

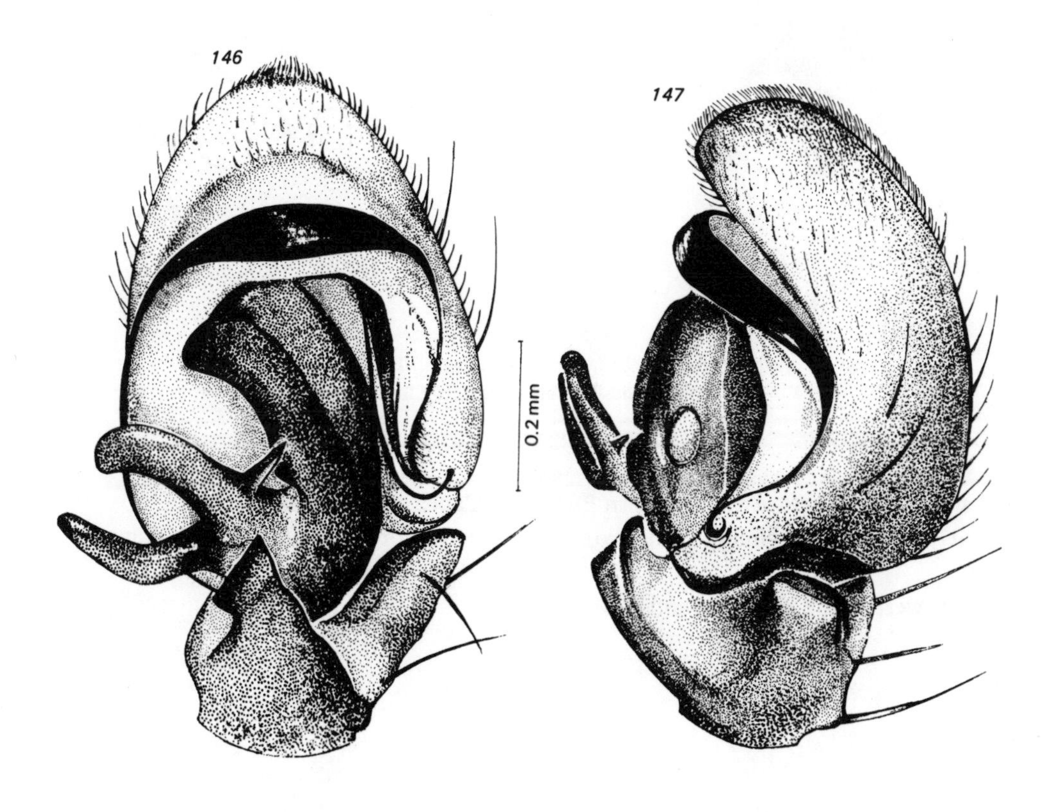

Figs. 146–147: ***Xysticus kochi*** **Thorell, 1872; male, left palpus**
146. ventral view; 147. lateral view

Tutaculum large, protruding inwards (Fig. 147). Tegulum with two large, thick, finger-like, curved apophyses forming a circle by bending towards each other (Fig. 146); back of shaft of upper apophysis with dark, upwards pointing spine (Figs. 146, 147). Embolar duct broad with tip slightly bent (Figs. 146, 147).

Female Epigynum: Central orifice divided by distinct, raised median septum (Fig. 148); basal rim of orifice protrudes markedly downwards above median septum (Fig. 148; dorsal-inner view, Fig. 149).

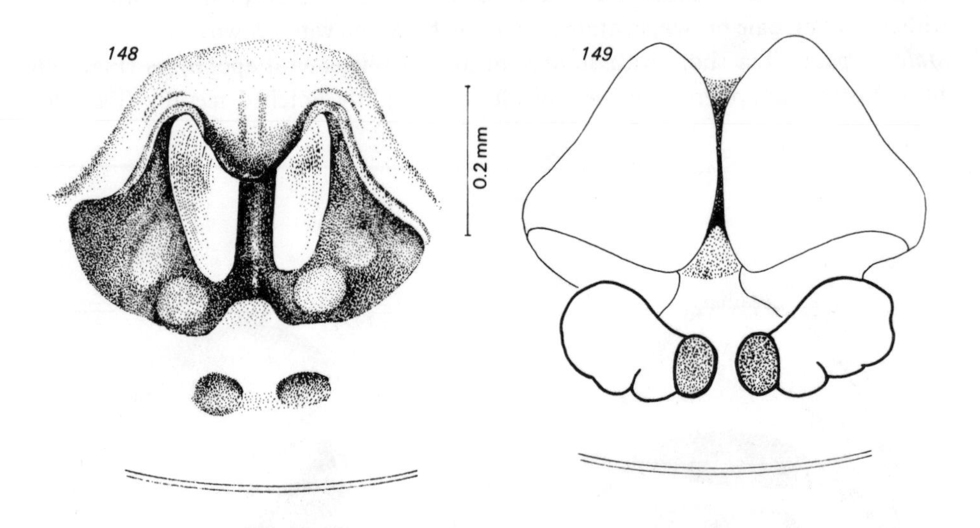

Figs. 148–149: *Xysticus kochi* Thorell, 1872; female
148. epigynum; 149. spermathecae, inner view

Distribution : North Africa, western Europe to Caucasus, Syria, Israel.
Israel: Very few specimens; only from the northern Jordan Valley (7) and Samaria (6).
Adults were found in March and in May.

Xysticus kempeleni Thorell, 1872
Figs. 150–153

Xysticus kempeleni Thorell, 1872, *Remarks on Synonyms of European Spiders*, Upsala, p. 245; Roewer, 1954, *Katalog der Araneae*, 2 (1): 902; Bonnet, 1959, *Bibliographia Araneorum*, 2 (5): 4879; Levy, 1976, *Israel J. Zool.*, 25: 25.

Thomisus graecus C. L. Koch, 1838. O. P.-Cambridge, 1872, *Proc. zool. Soc. Lond.*, 305 (misdetermination; not *Xysticus graecus* C. L. Koch, 1838).

Length of male 3.6–4.7 mm, female 4.8–7.1 mm. Coloration of prosoma brown with a wedge-shaped yellowish marking on back. Legs brown or mottled with white, mainly in females. Opisthosoma whitish with a light brown dentate pattern on back.

Male Palpus: Tibia with ventral, intermediate and retrolateral apophyses (Figs. 150, 151). Ventral apophysis curved inwards; tip round, slanting mesally with shaft slightly notched on mesal side (Fig. 150). Intermediate apophysis thick, cone-like, almost of same height as ventral apophysis (Figs. 150, 151). Retrolateral apophysis large, rounded on outside, with thick, sclerotic tip (Figs. 150, 151). Transparent tutaculum cup-like and rather large (Fig. 151). Tegulum with rounded central swelling turning gradually into an encircling, non-sclerotized, elevated structure, opening

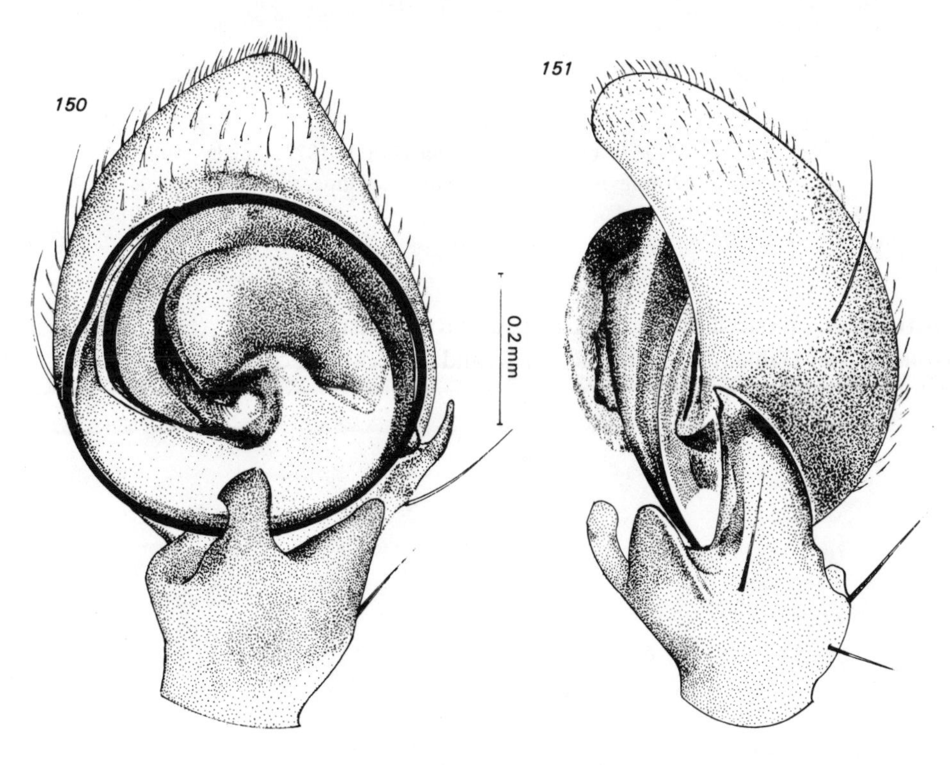

Figs. 150–151: *Xysticus kempeleni* Thorell, 1872; male, left palpus
150. ventral view; 151. lateral view

downwards at height of tutaculum or, occasionally, extending opposite ventral apophysis (Fig. 150).

Female Epigynum: Epigynal plate with slightly oblong central orifice, opening towards epigastric furrow in a shallow canal (Fig. 152). Central orifice partly surrounded by a double set of rims, the inner light-coloured and membranous, the outer dark and sclerotized (Fig. 152). On bottom of orifice, lines of two elongated loops are discerned (Fig. 152; dorsal-inner view, Fig. 153).

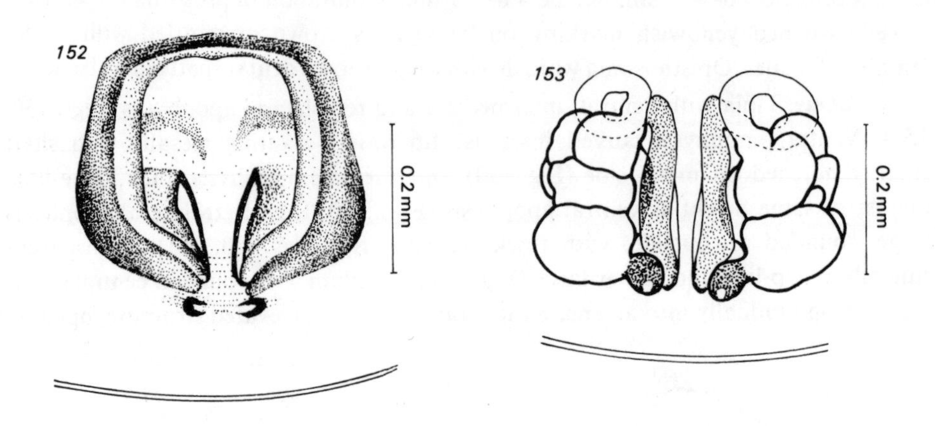

Figs. 152–153: *Xysticus kempeleni* Thorell, 1872; female
152. epigynum; 153. spermathecae, inner view

Distribution: Western Europe to Greece and Russia, Israel.
Israel: Upper Galilee (1), Carmel Ridge (3) and the Central Coastal Plain (8).
Adults are found in March.

Xysticus xerodermus Strand, 1913
Figs. 154–157

Xysticus xerodermus Strand, 1913, *Arch. Naturgesch.*, 79: 153; Roewer, 1954, *Katalog der Araneae*, 2 (1): 909; Bonnet, 1959, *Bibliographia Araneorum*, 2 (5): 4902; Levy, 1976, *Israel J. Zool.*, 25: 27.

Length of male 4.3–5.4 mm, female 5.9–9.6 mm. Coloration of prosoma reddish-brown mottled with black and with a dark, dorsal wedge-shaped marking enclosed by a yellowish band. Legs brown to black, mottled with white spots. Opisthosoma pale brown with a median dentate band on back.

Male Palpus: Tibia with ventral and retrolateral apophyses. Ventral apophysis extending away from bulb, slightly inclined inwards apically, with rounded swelling on inside (Figs. 154, 155). Retrolateral apophysis large, with thick, round base, the following part slender and sclerotic with tip pointing inwards (Fig. 155). Tutaculum cup-like and rather conspicuous (Figs. 154, 155). Tegulum, slightly off-centre, with a sclerotized, protruding crescent-shaped structure (Fig. 154).

Female Epigynum: Epigynal plate with almost round, central orifice, opening towards epigastric furrow (Fig. 156); septum-like elevation extending half-way into central orifice through its opening. Thick, sclerotized walls surround central orifice, turning into fine loops near opening (Fig. 156; dorsal-inner view, Fig. 157).

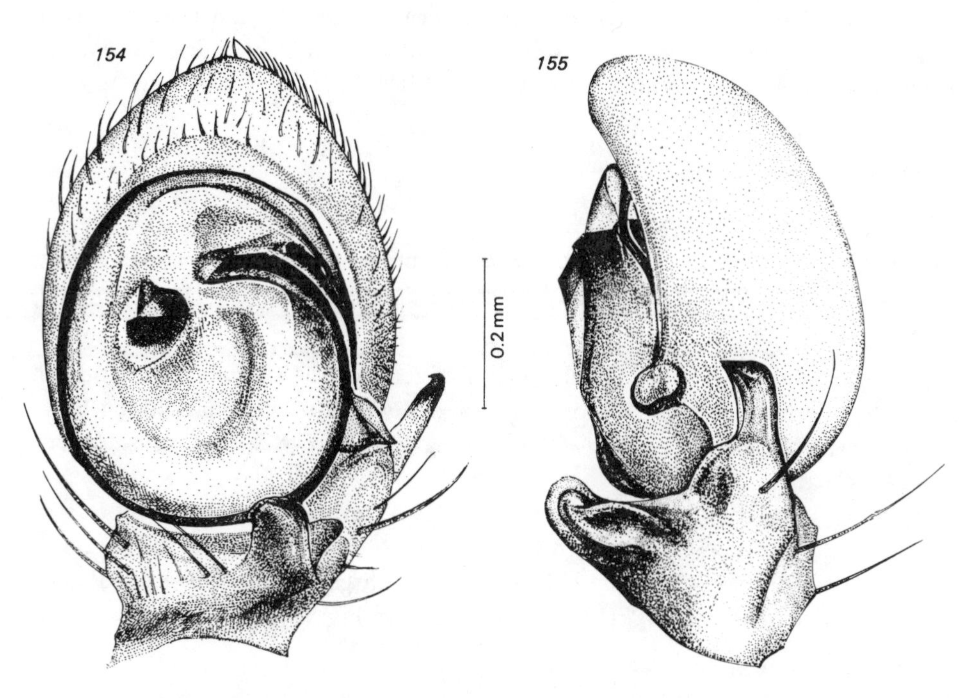

Figs. 154–155: *Xysticus xerodermus* Strand, 1913; male, left palpus
154. ventral view; 155. lateral view

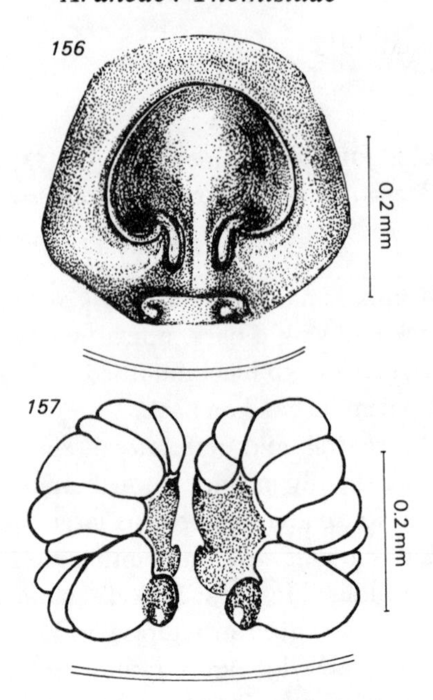

Figs. 156–157: *Xysticus xerodermus* Strand, 1913; female
156. epigynum; 157. spermathecae, inner view

Distribution, Israel: Throughout northern and central parts, from near Kinneret (7), Lower Galilee (2) to the Judean Desert (12) and the Foothills of Judea (Lahav, 10). Adults are found from November–December to May–June.

Xysticus graecus C. L. Koch, 1838
Figs. 158, 159

Xysticus graecus C. L. Koch, 1838, *Die Arachniden*, Nürnberg, IV, p. 65; Roewer, 1954, *Katalog der Araneae*, 2 (1): 901; Bonnet, 1958, *Bibliographia Araneorum*, 2 (4): 3792 (as *Proxysticus*); Levy, 1976, *Israel J. Zool.*, 25: 29.

Male as yet unknown from Israel. Length of female 7.3–9.6 mm. Coloration of prosoma yellowish to reddish-brown with a pale wedge-shaped marking on back. Legs brown mottled with white. Opisthosoma light brown with a pale-coloured dentate pattern.
Female Epigynum: Epigynal plate large, with long, tongue-like median protrusion projecting towards epigastric furrow (Fig. 158); protrusion slightly bilobed and raised at end, forming elongated hood-like cover. Large twisted folds extend along sides of median structure (Fig. 158; dorsal-inner view, Fig. 159).

Distribution: Hungary to Balkan countries, Israel.
Israel: Golan Heights (18).
Adult females were found in June–July.

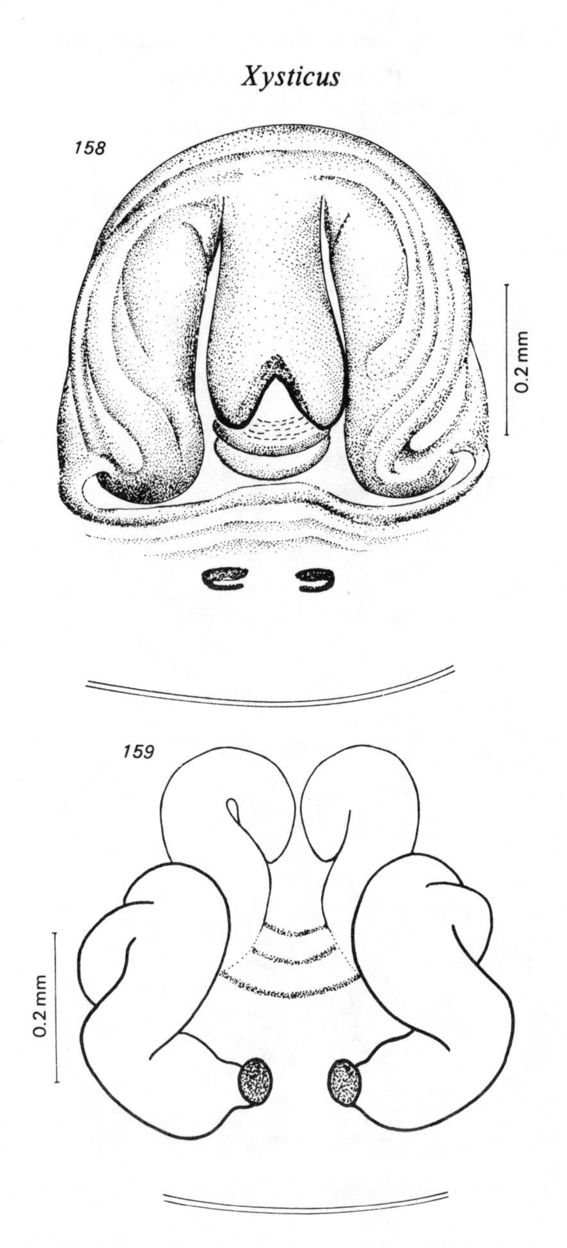

Figs. 158–159: *Xysticus graecus* C. L. Koch, 1838; female
158. epigynum; 159. spermathecae, inner view

Xysticus altitudinis Levy, 1976
Figs. 160, 161

Xysticus altitudinis Levy, 1976, *Israel J. Zool.*, 25: 31.

Male unknown. Length of female 10.3 mm. Coloration of prosoma and legs deep reddish-brown mottled with yellow markings. Opisthosoma yellowish-brown with a

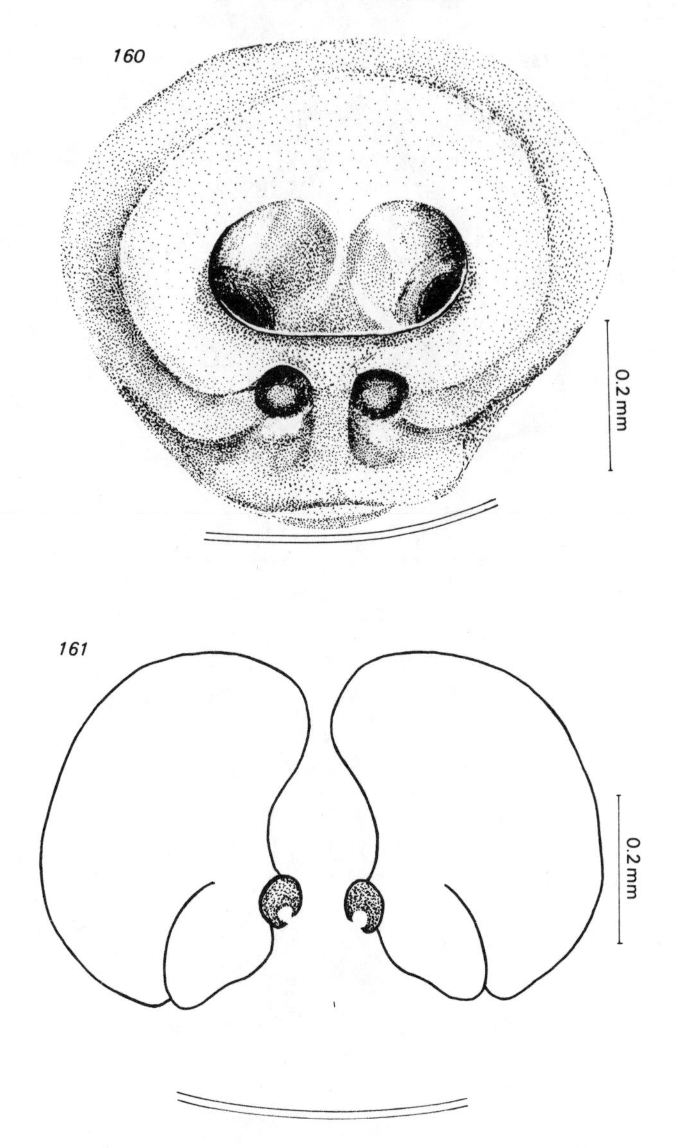

Figs. 160–161: *Xysticus altitudinis* Levy, 1976; female
160. epigynum; 161. spermathecae, inner view

light-coloured, dentate median band on back.

Female Epigynum: Epigynal plate relatively large. Almost round orifice situated centrally, bordered on side of epigastric furrow by distinct raised rim (Fig. 160); two dark openings are sunk below rim inside central orifice. Distal part of epigynum with two dark, rather conspicuous attachment points of the spermathecal apodemes (Fig. 160; dorsal-inner view, Fig. 161).

Distribution, Israel: Known only from Mt. Hermon, 2000 m (19); September.

Xysticus gymnocephalus Strand, 1915
Figs. 162, 163

Xysticus gymnocephalus Strand, 1915, *Arch. Naturgesch.*, 81: 148; Roewer, 1954, *Katalog der Araneae*, 2 (1): 901; Bonnet, 1959, *Bibliographia Araneorum*, 2 (5): 4878; Levy, *Israel J. Zool.*, 25: 32.

Male unknown. Length of female 5.4–8.7 mm. Coloration of prosoma yellowish-brown with a light-coloured wedge-shaped marking on back. Legs light brown mottled with white. Opisthosoma pale brown with a median, yellowish, dentate pattern.

Female Epigynum: Central part of epigynal plate slants gradually towards epigastric furrow (Fig. 162); distinct S-shaped rims enclose central part on each side. Elevated rims, on distal corners, show two dark sclerotic rings indicating attachment points of spermathecal apodemes (Fig. 162; dorsal-inner view, Fig. 163).

Distribution, Israel: Central Coastal Plain (8) and Judean Hills (11).
Adult females are found mainly from October–November to January.

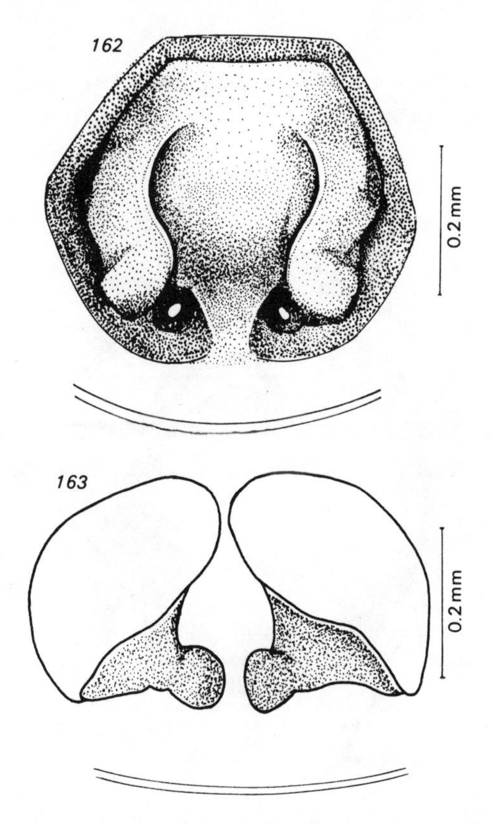

Figs. 162–163: *Xysticus gymnocephalus* Strand, 1915; female
162. epigynum; 163. spermathecae, inner view

Xysticus jezequeli Karol, 1966
Figs. 164, 165

Xysticus jezequeli Karol, 1966, *Commun. Fac. Sci. Univ. Ankara*, 11c: 1–5; Levy, 1976, *Israel J. Zool.*, 25: 33.

Length of male 4.4–4.8 mm; female unknown. Coloration of prosoma light to deep brown with a pale, wedge-shaped marking on back. Anterior pairs of legs dark brown, posterior pairs mottled with yellow spots. Opisthosoma yellowish-brown with a light dentate band, traversed longitudinally by a dark line.
Male Palpus: Tibia with ventral and retrolateral apophyses. Ventral apophysis strong, inclined laterally, with upper, inside corner protruding mesally (Fig. 164); upper, outer corner tapering slightly outwards (Fig. 165). Retrolateral apophysis thick, cone-like, more or less reaching height of ventral apophysis (Fig. 165). Tutaculum large, pointed, markedly projecting outwards, with deep concavity inside (Fig. 164). Tegulum smooth, convex, apically giving rise to thick, smooth embolar duct with tip slightly bent (Figs. 164, 165).

Distribution: Turkey, Israel.
Israel: Judean Hills (11).
Adult males are found from October to December. The matching female might be represented by *X. gymnocephalus*.

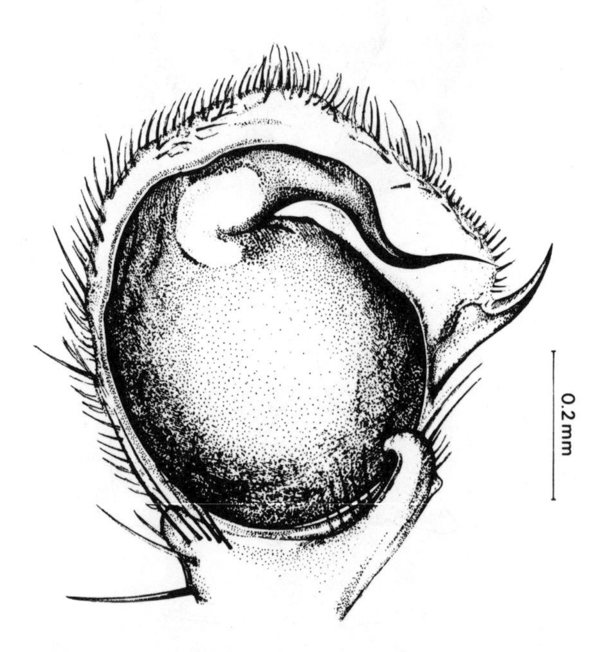

Fig. 164: *Xysticus jezequeli* Karol, 1966; male, left palpus, ventral view

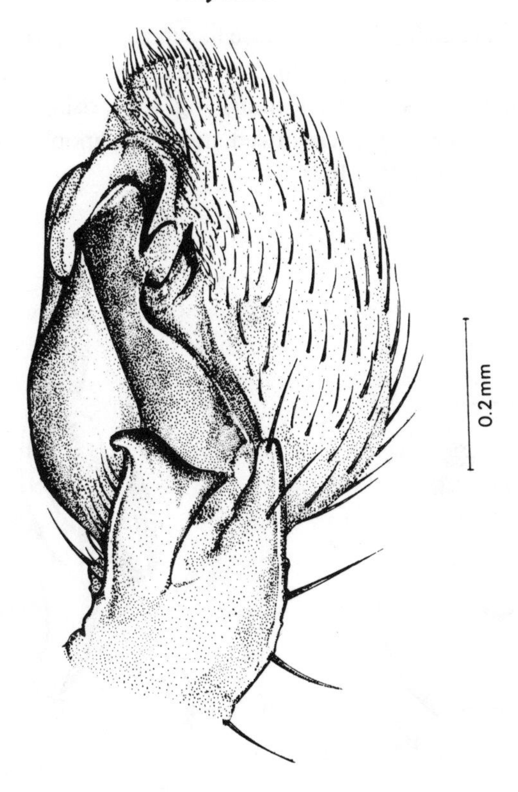

Fig. 165: *Xysticus jezequeli* Karol, 1966; male, left palpus, lateral view

Xysticus lalandei (Audouin, 1827)
Figs. 166–169

Thomisus lalandii Audouin, 1827, Explication sommaire des planches d'Arachnides, in: J. C. Savigny, *Description de l'Égypte*, Paris, 22: 398, pl. 6, fig. 12.
Xysticus lalandei —. Simon, 1885, *Exploration Scientifique de la Tunisie*, Paris, p. 16 : Roewer, 1954, *Katalog der Araneae*, 2 (1): 903; Levy, 1976, *Israel J. Zool.*, 25: 34.

Length of male 4.2 mm, female 9.5–11.9 mm. Coloration of prosoma yellowish-brown with a pale, dorsal, wedge-shaped marking; female also mottled with dark spots. Legs of female pale brown mottled with dark spots; in male, the tibiae are almost black. Opisthosoma whitish with a pale, dorsal, dentate pattern and a few darker spots on sides.
Male Palpus: Relatively small. Tibia with ventral, intermediate and retrolateral apophyses (Figs. 166, 167). Ventral apophysis apically wide and slightly inclined inwards (Figs. 166, 167). Intermediate apophysis low, rounded and broadly attached to retrolateral apophysis (Figs. 166, 167). Retrolateral apophysis wide at base, with gradually

tapering tip (Figs. 166, 167). Tegulum, close to centre, giving rise to fine, dark, sclerotized, encircling embolar duct (Fig. 166); fine, dark, projecting, crescent-shaped structure at centre, with upwards-turned opening (Fig. 166); rising apically on bulb, below tip of embolar duct, is a strongly sclerotized, hook-like apophysis with serrated upper edge and recurved tip (Figs. 166, 167).

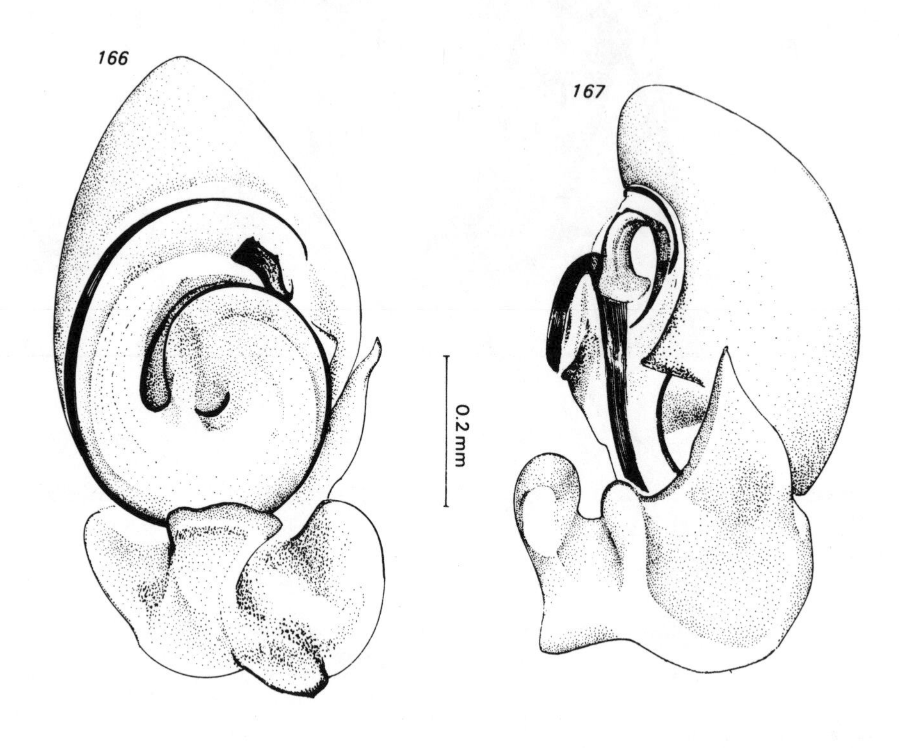

Figs. 166–167: *Xysticus lalandei* (Audouin, 1827); male, left palpus
166. ventral view; 167. lateral view

Female Epigynum: Epigynal plate on basal part with central, transparent, round, hood-like structure with triangular, tent-like opening (Fig. 168). Hood-like structure accompanied on each side by large, sclerotized lateral caps (Fig. 168). Distal part of epigynal plate traversed by elevated, dark rim, bordering two low, oval depressions (Fig. 168; dorsal-inner view, Fig. 169).

Distribution: Egypt, Sinai and probably North Africa.
Sinai: Southwestern Sinai (23).
Adults were found in October and January.

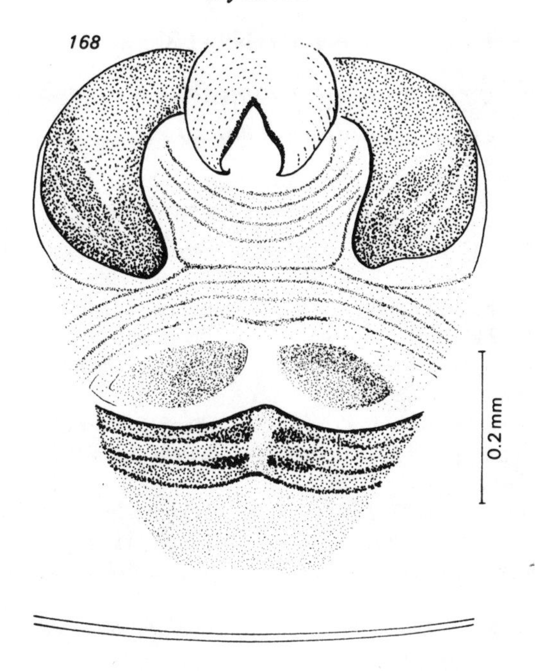

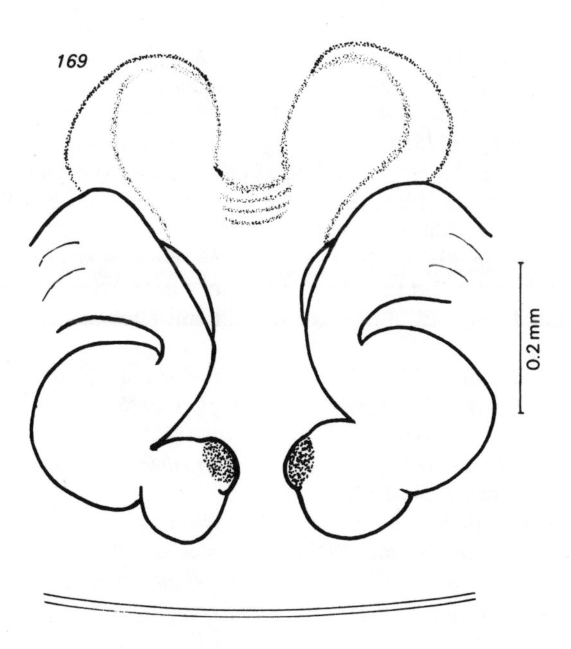

Figs. 168–169: *Xysticus lalandei* (Audouin, 1827); female
168. epigynum; 169. spermathecae, inner view

APPENDIX

List of Emendations of Thomisidae Formerly Reported Mainly from Israel

Former Name	*Emendation*
acerbus (Xysticus) Thorell, 1872. Bodenheimer, 1937: 241	*xerodermus (Xysticus)* Strand, 1913
aharonii (Oxyptila) Strand, 1913: 150. Bodenheimer, 1937: 240	nomen dubium (immature ♀ *Xysticus*, unrecognizable)
albus (Thomisus) Gmelin, 1789. Strand, 1913: 150; Bodenheimer, 1937: 240	*onustus (Thomisus)* Walckenaer, 1805
audouini (Synaema) Roewer, 1951: 449	*diana (Synaema)* (Audouin, 1827)
blitea (Oxyptila) Simon, 1875: 236. Bodenheimer, 1937: 240	*bliteus (Xysticus)* (Simon, 1875)
buffonii (Thomisus) Audouin, 1827: 396	*buffoni (Heriaeus)* (Audouin, 1827)
bufo (Thomiseus) Dufour, 1820: 206	? *bufo (Xysticus)* (Dufour, 1820)
calycina (Misumena) Linnaeus, 1758. Bodenheimer, 1937: 240	*onustus (Thomisus)* Walckenaer, 1805
cerina (Runcinia) (C. L. Koch, 1845). Roewer, 1954: 853	*lateralis (Runcinia)* (C. L. Koch, 1838)
claviger (Thomisus) O. P.-Cambridge, 1872: 306	*clavigera (Oxyptila)* (O. P.-Cambridge, 1872)
confluens (Xysticus) C. L. Koch, 1845. O. P.-Cambridge, 1872: 305; Bodenheimer, 1937, 240	*tricoloripes (Oxyptila)* Strand, 1913
cristatus (Araneus) Clerck, 1757: 136	*cristatus (Xysticus)* (Clerck, 1757)
cristatus (Xysticus) (Clerck, 1757). Pavesi, 1895: 8	*edax (Xysticus)* (O. P.-Cambridge, 1872)
diana (Thomisus) Audouin, 1827: 399	*diana (Synaema)* (Audouin, 1827)
edax (Thomisus) O. P.-Cambridge, 1872: 303	*edax (Xysticus)* (O. P.-Cambridge, 1872)
globosa (Aranea) Fabricius, 1775: 432	*globosum (Synaema)* (Fabricius, 1775)
globosa (Diaea) (Fabricius, 1775), Pavesi, 1895: 8	*globosum (Synaema)* (Fabricius, 1775)
graecus (Xysticus) C. L. Koch, 1838. O. P.-Cambridge, 1872: 305; Bodenheimer, 1937: 241	*kempeleni (Xysticus)* Thorell, 1872
hirtus (Thomisus) Audouin, 1827: 397	*hirta (Oxyptila)* (Audouin, 1827)
hirtus (Xysticus) O. P.-Cambridge, 1876: 581	*bliteus (Xysticus)* (Simon, 1875)
jaffa (Oxyptila) Strand, 1915: 151. Bodenheimer, 1937: 240	nomen dubium (immature ♀ *Oxyptila*, unrecognizable)
lalandii (Thomisus) Audouin, 1827: 398	*lalandei (Xysticus)* (Audouin, 1827)
ornatum (Synaema) Chyzer & Kulczyński, 1891: 86	*plorator (Synaema)* (O. P.-Cambridge, 1872)
paradoxus (Monaeses) (Lucas, 1846). Strand, 1915: 147; Bodenheimer, 1937: 240	*israeliensis (Monaeses)* Levy, 1973
pickardi (Oxyptila) Levy, 1975	*tricoloripes (Oxyptila)* Strand, 1913
piochardi judaorum (Tmarus) Strand, 1915: 147	*piochardi (Tmarus)* (Simon, 1866)
plorator (Thomisus) O. P.-Cambridge, 1872: 306	*plorator (Synaema)* (O. P.-Cambridge, 1872)
promiscuus (Xysticus) O. P.-Cambridge, 1876: 581	*cristatus (Xysticus)* (Clerck, 1757)
rectilineus (Thomisus) O. P.-Cambridge, 1872: 306	*rectilineus (Xysticus)* (O. P.-Cambridge, 1872)

? *richteri (Synaema)* Utochkin, 1960: 1022	*plorator (Synaema)* (O. P.-Cambridge, 1872)
rigida (Oxyptila) (O. P.-Cambridge, 1872). Kulczyński, 1908: 70	*tricoloripes (Oxyptila)* Strand, 1913
rigidus (Thomisus) O. P.-Cambridge, 1872: 305	*rigida (Oxyptila)* (O. P.-Cambridge, 1872)
rotundatus (Thomisus) Walckenaer, 1805. O. P.-Cambridge, 1872: 306	*globosum (Synaema)* (Fabricius, 1775)
setiger (Thomisus) O. P.-Cambridge, 1872: 307	*setiger (Heriaeus)* (O. P.-Cambridge, 1872)
similis (Xysticus) Kulczyński, 1911: 33. Bodenheimer, 1937: 241	*ferus (Xysticus)* O. P.-Cambridge, 1876
spinifer (Thomisus) O. P.-Cambridge, 1872: 308. Bodenheimer, 1937: 240	*citrinellus (Thomisus)* Simon, 1875
tristrami (Thomisus) O. P.-Cambridge, 1872: 304	*tristrami (Xysticus)* (O. P.-Cambridge, 1872)
tristrami (Xysticus) (O. P.-Cambridge, 1872). Pavesi, 1895: 8	*ferus (Xysticus)* O. P.-Cambridge, 1876 in part, *rectilineus (Xysticus)* (O. P.-Cambridge, 1872) in part
tristrami (Xysticus) (O. P.-Cambridge, 1872). Kulczyński, 1908: 73	*ferus (Xysticus)* O. P.-Cambridge, 1876
truncata (Misumena) (Pallas, 1772). Pavesi, 1895: 8	*onustus (Thomisus)* Walckenaer, 1805
truncatus (Pistius) (Pallas, 1772). Bodenheimer, 1937: 240	*onustus (Thomisus)* Walckenaer, 1805
unica (Oxyptila) Strand, 1915: 150. Bodenheimer, 1937: 240	nomen dubium (immature ♀ *Oxyptila*, unrecognizable)
varius (Thomisus) O. P.-Cambridge, 1872: 303. Bodenheimer, 1937: 241	nomen dubium (immature ♀ *Xysticus*, unrecognizable)
vatia (Misumena) (Clerck, 1757). Pavesi, 1895: 8	*onustus (Thomisus)* Walckenaer, 1805
viaticus (Xysticus) C. L. Koch, 1835. Bodenheimer, 1937: 241	*kochi (Xysticus)* Thorell, 1872

REFERENCES*

Audouin, V. (1827) 'Explication sommaire des planches d'Arachnides de l'Égypte et de la Syrie', in: *Description de l'Égypte*, J. C. Savigny (ed.), Paris, 22: 291–430.

Bodenheimer, F. S. (1937) 'Prodromus Faunae Palestinae', *Mémoires Inst. Égypte*, 33: 1–286.

Bonnet, P. (1945–1961) *Bibliographia Araneorum*, Toulouse, Impre. Douladoure, 7 vols.

Cambridge, O. P.- (1872) 'General list of the spiders of Palestine and Syria, with descriptions of numerous new species and characters of two new genera', *Proc. zool. Soc. Lond.*, pp. 212–354.

— (1876) 'Catalogue of a collection of spiders made in Egypt, with descriptions of new species and characters of a new genus', *Proc. zool. Soc. Lond.*, pp. 541–630.

Fet, V. Y. (1982), 'New for the USSR fauna — A spider from Kopetdagh *Oxyptila tricoloripes* Strand, 1913, (Aranei, Thomisidae)', *Trans. Acad. Sci. Turkmen. SSR*, 5: 74–75 (in Russian).

Kulczyński, V. (1908) 'Fragmenta arachnologica VI; X, Araneae nonnullae in Cypro insula et in Palaestina a cel. Prof. Dr. G. Cecconi lectae', *Bull. Acad. Sci. Cracovie*, pp. 49–86.

— (1911) 'Fragmenta arachnologica IX; XVI, Aranearum species nonnullae in Syria a Rev. P. Bovier-Lapierre et in Palaestina a Rev. E. Schmitz collectae', *Bull. Acad. Sci. Cracovie*, pp. 12–55.

Levy, G. (1970) 'The life cycle of *Thomisus onustus* (Thomisidae, Araneae) and outlines for the classification of the life histories of spiders', *J. Zool. Lond.*, 160 (4): 523–536.

— (1973) 'Crab-spiders of six genera from Israel (Araneae: Thomisidae)', *Israel J. Zool.*, 22: 107–141.

— (1975) 'The spider genera *Synaema* and *Oxyptila* in Israel (Araneae: Thomisidae)', *Israel J. Zool.*, 24: 155–175.

— (1976) 'The spider genus *Xysticus* (Araneae: Thomisidae) in Israel', *Israel J. Zool.*, 25: 1–37.

Pavesi, P. (1895) 'Viaggio del Dott. E. Festa in Palestina, nel Libano e regioni vicine. XIV. Aracnidi', *Boll. Musei Zool. Anat. Comp. R. Univ. Torino*, 10 (216): 1–11.

Roewer, C. F. (1954) *Katalog der Araneae*, Brussels, 2 (1): 1–923.

Simon, E. (1892) 'Liste des arachnides recueillis en Syrie par M. le Dr. Théod. Barrois', *Rev. Biol. Nord Fr., Lille*, 5: 80–84.

Strand, E. (1913) 'Erste Mitteilung über Spinnen aus Palästina, gesammelt von Herrn Dr. J. Aharoni', *Arch. Naturgesch. Berlin*, 79 (A, 10): 147–162.

— (1915) 'Dritte Mitteilung über Spinnen aus Palästina, gesammelt von Herrn Dr. J. Aharoni', *Arch. Naturgesch. Berlin*, 81 (A; 2): 134–171.

* Listed are only selected references on the thomisid spiders of Israel; complete lists of references are given in the publications by G. Levy.

INDEX

Synonyms in italics. The principle reference to each valid name in bold type.

abbreviatus Walckenaer, Thomisus 37
acerbus Thorell, Xysticus 108
aharonii Strand, *Oxyptila* 108
albus Gmelin, Thomisus 108
altitudinis Levy, Xysticus 75, **101–102**
aradensis Levy, Oxyptila 61, **69–70**
audouini Roewer, Synaema 58, 108

bidentatus Kulczyński, Thomisus 36, **41–42**
blitea Simon, *Oxyptila* 86, 108
bliteus (Simon), Xysticus 73, 74, **86–87**, 108
bragantinus Capello, *Thomisus* 31
brevipes Hahn, *Thomisus* 60
buffoni (Audouin), Heriaeus 47, **51–52**, 108
buffonii Audouin, *Thomisus* 51, 108
bufo Dufour, *Thomiseus* 88, 108
?bufo (Dufour), Xysticus 74, **88–89**, 108

calycina Linnaeus, *Misumena* 108
caperatoides Levy, Xysticus 74, **83**
caperatus Simon, Xysticus 74, 75, **84–86**
cerina (C.L. Koch), Runcinia 45, 108
citrinellus Simon, Thomisus 36, **39–40**, 109
claviger O.P.-Cambridge, *Thomisus* 65, 108
clavigera (O.P.-Cambridge), Oxyptila 1, 62, **65**, 108
complicata Levy, Oxyptila 61, **66–67**
confluens (C.L. Koch), *Thomisus* 62
confluens C.L. Koch, *Xysticus* 108
cristatus Clerck, *Araneus* 71, 90, 108
cristatus (Clerck), Xysticus 74, 75, **90–92**, 108
cristatus (Clerck), Xysticus 92, 108

dewitzi Simon, Firmicus **33–34**
diana (Audouin), Synaema 53, **58–59**, 108
diana Audouin, *Thomisus* 58, 108

edax O.P.-Cambridge, *Thomisus* 92, 108
edax (O.P.-Cambridge), Xysticus 74, 75, **92–94**, 108

ferus O.P.-Cambridge, Xysticus 75, **81–82**, 83, 109
Firmicus Simon 17, **31–32**

globosa Fabricius, *Aranea* 53, 55, 108
globosa (Fabricius), *Diaea* 108
globosum (Fabricius), Synaema 53, **55–56**, 108, 109
graecus C.L. Koch, *Thomisus* 97
graecus C.L. Koch, Xysticus 75, **100–101**
graecus C.L. Koch, Xysticus 108
gymnocephalus Strand, Xysticus 75, **103**, 104

hazevensis Levy, Tmarus 22, **29–30**
Heriaeus Simon 13, 18, **47–48**
hirta (Audouin), Oxyptila 108
hirtus Audouin, *Thomisus* 108
hirtus Latreille, *Thomisus* 47
hirtus O.P.-Cambridge, Xysticus 86, 108

israeliensis Levy, Monaeses **20–21**, 108

jaffa Strand, *Oxyptila* 108
jezequeli Karol, Xysticus 74, **104–105**
judaea Levy, Oxyptila 62, **68**, 69

kempeleni Thorell, Xysticus 74, 75, **97–98**, 108
kochi Thorell, Xysticus 74, 75, **95–96**, 109

lalandei (Audouin), Xysticus 74, 75, **105–107**, 108
lalandii Audouin, *Thomisus* 105, 108
lateralis (C.L. Koch), Runcinia **45–46**, 108
lateralis C.L. Koch, *Thomisus* 43, 45

Misumeninae 17
Monaeses Thorell 17, **18–19**, 22
Monastes Lucas 18

omega Levy, Oxyptila 62, **70–71**
onustus Walckenaer, Thomisus 13, 14, 15, 35, 36, **37–38**, 108, 109
ornatum Chyzer & Kulczyński, *Synaema* 108
Oxyptila Simon 13, 14, 15, 17, 18, **60–62**, 72, 108, 109
Ozyptila Simon 60

paradoxus Lucas, *Monastes* 18
paradoxus (Lucas), Monaeses 20, 108
peronii Audouin, Thomisus 37
pickardi Levy, Oxyptila 62, 108
pigra Walckenaer, *Aranea* 22
piochardi judaorum Strand, Tmarus 25, 108
piochardi Simon, *Thomisus* 25
piochardi (Simon), Tmarus 22, **25–27**
plorator (O.P.-Cambridge), Synaema 53, 55, **56–58**, 108, 109
plorator O.P.-Cambridge, *Thomisus* 56
promiscuus O.P.-Cambridge, Xysticus 90, 108
Proxysticus Dalmas 84, 88, 100

rectilineus O.P.-Cambridge, *Thomisus* 76, 108
rectilineus (O.P.-Cambridge), Xysticus 73, 75, **76–78**, 108, 109
?*richteri* Utochkin, Synaema 109
rigida (O.P.-Cambridge), Oxyptila 62, **65–66**, 67, 109
rigida (O.P.-Cambridge), Oxyptila 62, 109
rigidus O.P.-Cambridge, *Thomisus* 65, 109
rotundatus Walckenaer, *Thomisus* 109
Runcinia Simon 13, 14, 17, **43–44**

setiger (O.P.-Cambridge), Heriaeus 47, **49–50**, 109
setiger O.P.-Cambridge, *Thomisus* 49, 109
similis Kulczyński, *Xysticus* 81, 109
spinifer O.P.-Cambridge, *Thomisus* 39, 109
Synaema Simon 13, 14, 17, 18, **53–55**
Synema Simon 53

Thomisus Walckenaer 13, 14, 17, **35–36**
Tmarus Simon 17, **22–24**
tricoloripes Strand, Oxyptila 61, **62–64**, 108, 109
tristrami O.P.-Cambridge, *Thomisus* 79, 109
tristrami (O.P.-Cambridge), Xysticus 73, 75, **79–81**, 109
tristrami (O.P.-Cambridge), Xysticus 76, 81, 109
truncata (Pallas), *Misumena* 37, 109
truncatus (Pallas), *Pistius* 37, 109

unica Strand, *Oxyptila* 109

varius O.P.-Cambridge, *Thomisus* 109
vatia (Clerck), Misumena 13
vatia (Clerck), *Misumena* 37, 109
viaticus C.L. Koch, Xysticus 95, 109

xerodermus Strand, Xysticus 74, 76, **99–100**, 108
Xysticus C.L. Koch 13, 14, 15, 17, 18, **71–76**, 77, 108, 109

yerohamus Levy, Tmarus 22, **28–29**

MAP

Geographical Areas in Israel and Sinai

KEY

1. Upper Galilee
2. Lower Galilee
3. Carmel Ridge
4. Northern Coastal Plain
5. Valley of Yizre'el
6. Samaria
7. Jordan Valley and Southern Golan
8. Central Coastal Plain
9. Southern Coastal Plain
10. Foothills of Judea
11. Judean Hills
12. Judean Desert
13. Dead Sea Area
14. 'Arava Valley
15. Northern Negev
16. Southern Negev
17. Central Negev
18. Golan Heights
19. Mount Hermon
20. Northern Sinai
21. Central Sinai Foothills
22. Sinai Mountains
23. Southwestern Sinai

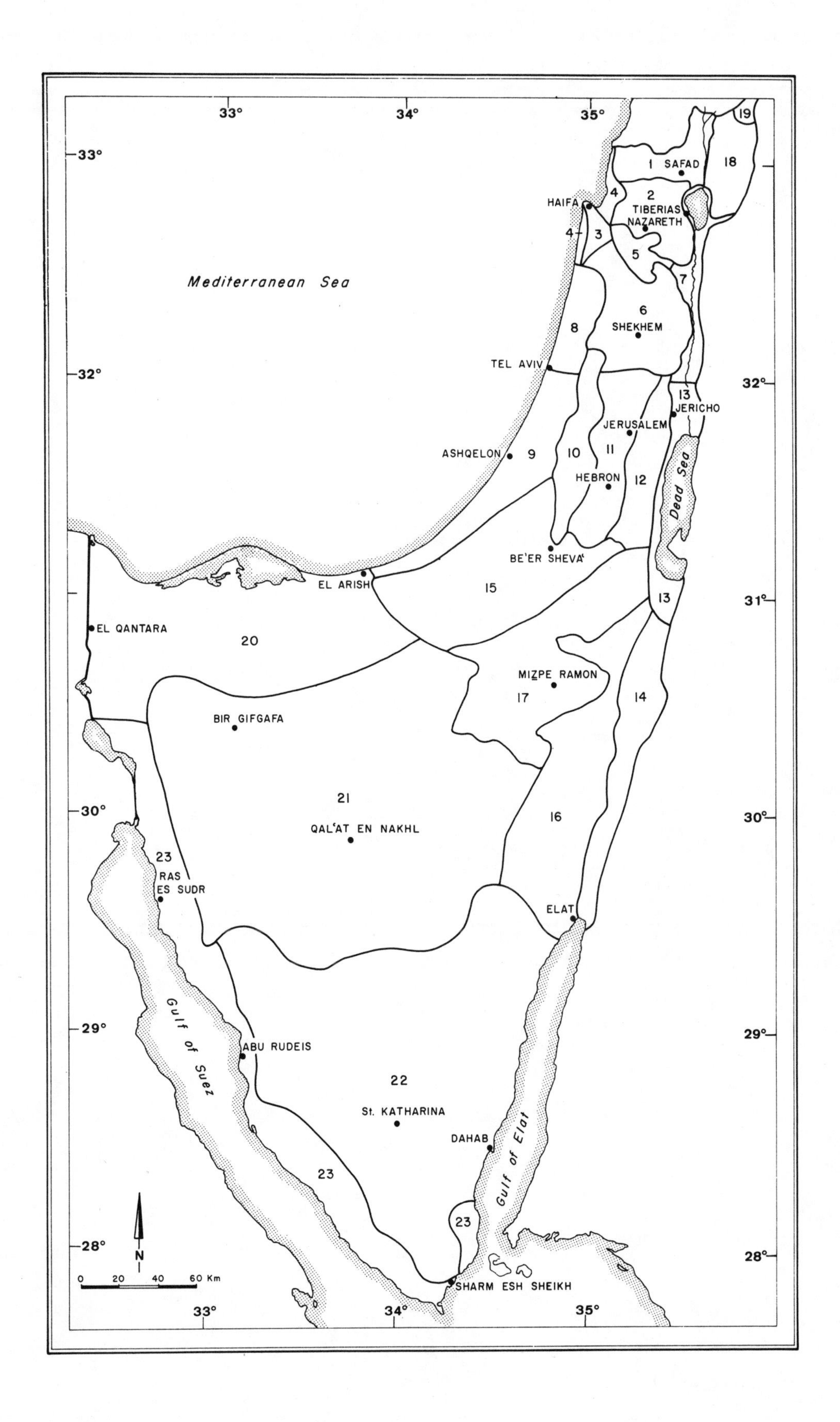

33°
34°
35°
33°
32°
31°
30°
29°
28°
Mediterranean Sea
1 SAFAD
18
19
2
TIBERIAS
NAZARETH
HAIFA
4
3
5
7
6
SHEKHEM
8
TEL AVIV
13
JERICHO
JERUSALEM
ASHQELON
9
10
11
HEBRON
12
Dead Sea
BE'ER SHEVA'
EL ARISH
15
13
EL QANTARA
20
MIZPE RAMON
17
14
BIR GIFGAFA
21
QAL'AT EN NAKHL
16
23
RAS
ES SUDR
ELAT
Gulf of Suez
ABU RUDEIS
22
St. KATHARINA
DAHAB
Gulf of Elat
23
23
N
0 20 40 60 Km
SHARM ESH SHEIKH

נדפס בישראל
תשמ"ה

כתבי האקדמיה הלאומית הישראלית למדעים
החטיבה למדעי־הטבע

החי של ארץ־ישראל

עכבישנים 2 : עכבישאים : עכבישים סרטניים

(ARANEAE: THOMISIDAE)

מאת

גרשֹם לוי

ירושלים תשמ״ה